THE BOY
ELECTRICIAN

Alfred P. Morgan

E P B M

ECHO POINT BOOKS & MEDIA, LLC

The Boy Electrician
Third edition, reprinted
Written and illustrated by Alfred P. Morgan

ISBN: 978-1-62654-981-4

Published by Echo Point Books & Media
www.EchoPointBooks.com

Original Copyright © 1913, 1929, 1940 Lathrop, Lee &
Shepard Co.

Cover layout by Adrienne Nunez, Echo Point Books & Media

Printed in the U.S.A.

IMPORTANT TO READ WARNING:

While this book has stood the test of time as a clear and fun guide to understanding electricity, some of its practices are dated and no longer considered safe. All of the experiments included in this text should only be completed under adult supervision, and some of these experiments **should not be attempted.**

The publisher strongly recommends that the following experiments never be undertaken:

> •The X-ray machine. The X-ray machine project is included in this printing for the intellectual understanding it offers the reader. **DO NOT** attempt to build or use an X-ray machine.

> •Amalgamation of zinc with mercury. Contemporary safety standards recommend avoiding the handling of mercury, which is a retentive neurotoxin that can be absorbed through the skin and/or enter the body as vapor. **DO NOT** work with mercury. The same safety standards apply to lead, which is also a retentive neurotoxin. **DO NOT** work with lead.

We have marked some pages in the book indicating warnings, but in general all safety issues are the responsibility of the reader as noted below.

A General Warning:

The material in this book is offered for information purposes only. While the publisher believes only a small number of the experiments included in these pages pose some possible danger, we urge caution and adult supervision for all experiments undertaken. This book was written in a different era when the health risks from certain materials and procedures were unknown. The reader is responsible for their own safety. The publisher will not be held liable for any harm resulting from experiments attempted in this book. The reader should proceed at their own risk.

PREFACE

ONCE upon a time, and this is a true tale, a boy had a whole railroad system for a toy. The train ran automatically, propelled by tiny electric motors, the signals went up and down, the station was reached, a bell rang, the train moved on again and was off on its journey around many feet of track to come back over the old route.

The boy viewed his gift with raptured eyes, and then his face changed and he cried out in the bitterness of his disappointment: "But what do I do?" The toy was so elaborate that the boy was left entirely out of the play. Of course he did not like it. His cry tells a long story.

The prime instinct of almost any boy at play is to *make* and to *create*. He will *make* things of such materials as he has at hand, and use the whole force of dream and fancy to *create* something out of nothing. The five-year-old will lay half a dozen wooden blocks together with a spool on one end and tell you it is a steam train. And it is. He has both made and created an engine, which he sees but which you don't, for the blocks and spool are only a symbol of his creation. Give his older brother a telephone receiver, some wire and bits of brass and he will make a radio outfit and listen to music and messages sent through space from sources hundreds of miles away. The radio set is not a symbol. It is something real. It can be heard and seen in operation. And as soon as the mystery of this modern wonder more firmly grips your imagination, you perhaps may come to realize that we are living more and more in the age of electricity. Electricity propels our trains, lights our houses and streets, carries our messages, makes our clothes, cures our ills, warms us, cooks for us and performs an innumerable number of other tasks at the turning of a little switch. A mere list is impossible.

Almost every boy experiments at one time or another with electricity and electrical apparatus. It is my purpose in writing this book to open this wonderland of science and present it in a manner which can be readily understood, and wherein a boy can "do something."

The apparatus and experiments that I have described have been constructed and carried out by *boys*. Their problems and their questions have been studied and remedied. I have tried to present practical matter considered wholly from a boy's standpoint, and to show the young experimenter just what he can do with the tools and materials in his possession or not hard to obtain.

To the boy interested in science, a wide field is open. There is no better education for any boy than to begin at the bottom of the ladder and climb the rungs of scientific knowledge, step by step. It equips him with information which may prove of inestimable worth.

New developments in science will never cease. Invention will follow invention. The unexpected is often a valuable clue. The Edisons and Teslas have not discovered everything. Inspiration is but the starting-point. Success means work—days, nights, weeks, and years of it—but more than that it means an overwhelming curiosity and indefatigable persistance.

There can be no boy who will follow exactly any directions given to him, or do exactly as he is told, of his own free will. He will "bolt," at the first opportunity. If forced or obliged to do as he is directed, his action will be accompanied by a host of "whys." Therefore in presenting the following chapters I have not only told how to *make* the various motors, telegraphs, telephones, batteries, etc., but have also explained the principles of electricity upon which they depend for their operation, and how the same thing is accomplished in the everyday world. In giving directions or in making cautions, I have usually given the reason for so doing in the hope that this information may be a stimulant to the imagination of the young experimenter and a useful guide in enabling him to proceed along some of the strange roads on which he will surely go.

ALFRED P. MORGAN

Upper Montclair, N. J.
March 15, 1940.

CONTENTS

CHAPTER I

CHAPTER II

CHAPTER III

Chapter XV

Chapter XVI

Chapter XVII

CHAPTER I

MAGNETS AND MAGNETISM

IF George Washington, or even Abraham Lincoln, could become alive today and visit one of our large cities, he would not know the words to name some of the things he would see. Elevators, subway trains, electric lights, radios, automobiles, etc., did not exist and their names were not part of the English language when either of these two famous Americans was alive.

Science brings new words into our language and it constantly changes our mode of living. Electrical science probably has made the greatest changes in our world within the last fifty years. It has made us more comfortable, healthier, made our days longer, given us many pleasures. It has brought us the telephone, radio, electric lights, motors, sound pictures, television, new materials, medicines, and a host of other things.

And all of these wonders have been invented and perfected by men who did not know what electricity is.

No one knows what electricity is. There have been many *theories* or *attempts* to explain what this mysterious force may actually be, but all of them have been mere guesses and cannot be proven.

It is not necessary to know what electricity is in order to use it and make it obey, or even in order to invent new electrical devices. It is merely necessary to know and understand the laws which govern electricity's actions and behavior.

There is nothing mysterious about electricity's actions. It

always will do exactly the same thing under exactly the same circumstances.

In order to understand electricity's behavior it is necessary to know something about *magnetism*.

THE DISCOVERY OF MAGNETISM

Over two thousand years ago, in far-away Asia Minor, a shepherd, guarding his flocks on the slope of Mount Ida, suddenly found the iron-shod end of his staff adhering to a stone. The shepherd's name was Magnes. Upon looking farther around about him Magnes found many other pieces of this hard black mineral, the smaller bits of which tended to cling to the iron nails and studs in the soles of his sandals.

This mineral, which was an ore of iron, consisting of iron and oxygen, became known as the "Magnesstone" or Magnet. The Greeks called it magnetite. As masses of the magnetic ore were discovered in various parts of the world,

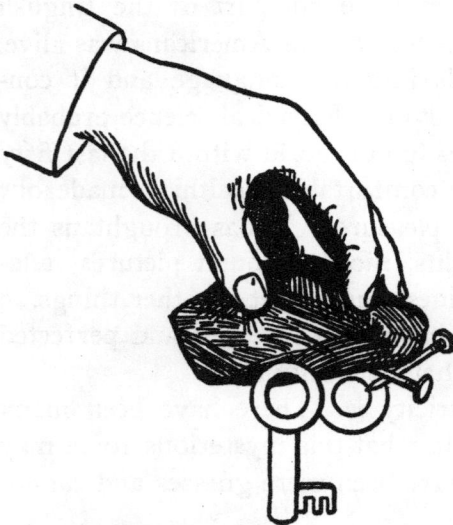

FIG. 1. — A Natural Magnet or Lodestone Will Attract Bits of Iron and Steel.

the stories of its attractive power became greatly exaggerated, especially during the middle ages. "Magnetic mountains" which would pull iron nails out of ships was one of the supposed terrors of the sea until nearly the eighteenth century. A magnet stone, it was believed, if carried in the pocket, would cure gout, headache, cramps, prevent baldness, and draw poisons from wounds. Such stories were just a lot of pure nonsense.

The story of Magnes, the shepherd, who first found the magnet is only a legend. There are many legends about the discovery of the magnet. The truth is that no one knows when the mineral which attracts iron objects or bits of iron ore was first noticed. After all, that knowledge, while it would be interesting, would be of no real importance.

The First Magnetic Compass

It is much more important that some one discovered that a magnet-stone hung by a thread in a suitable manner always will tend to point North and South.

A simple bit of lodestone suspended by a thread was the fore-runner of the modern compass. So the "Magnes-stone" also became called the "lodestone" or "leading stone." It was of great value to the ancient navigators, who, up to that time, had hardly dared venture out of sight of land because it was often difficult to find the way back. Since the lodestone would point north and south, here was a means of telling direction. A ship could now be steered in cloudy

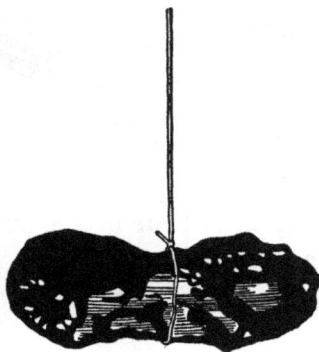

FIG. 2. — A Piece of Lodestone Suspended by a Thread Was Probably the First Compass.

weather when the sun was obscured and on nights when the pole-star could not be seen.

It is not known who made the first compass or when it was first used. The Chinese are sometimes given credit for its invention but there is abundant evidence that it was first used in Europe.

The most amazing thing about a bit of lodestone is its ability to impart its strange power to a piece of steel. If a piece of hardened steel is rubbed on a lodestone, *it becomes a magnet.*

The first mariner's compasses were called *gnomons.* They consisted of a steel needle which had been rubbed upon a lodestone until it became a magnet. Then the needle was thrust through a reed or a short piece of wood which floated on the surface of a vessel of water. It was given a circular movement in the water, and as soon as it came to rest, the point on the horizon to which the north-seeking end of the needle pointed was noted carefully and the ship's course laid accordingly.

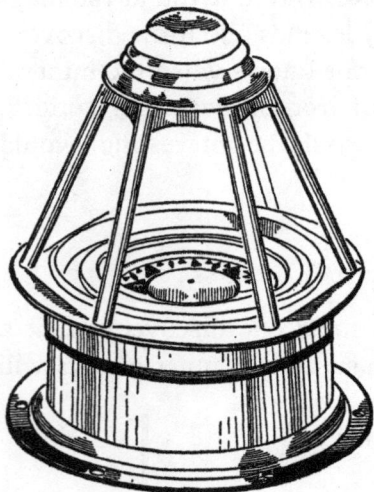

FIG. 3.—A Mariner's Compass Enclosed in Its Binnacle or Case.

MODERN COMPASSES

The little pocket compasses which the surveyors and hunters use to tell direction on land are as simple as the old gnomon. A flat steel needle which has been magnetised is supported on a pivot at its center. The needle swings over a scale divided into degrees and marked with the points of the compass. The needle and its scale are enclosed in a small watch-shaped case.

Naval vessels and many commercial ships are now equipped

with gyro compasses. These instruments are driven by an electric motor. They are not governed by the earth's magnetism but by the rotation of the earth.

The modern magnetic mariner's compass, used for steering ships, is constructed differently from a pocket compass. It consists of three parts, called the *bowl,* the *card* and the *needle.*

The bowl, which contains the card and needle, is usually a hemispherical brass receptacle, suspended in a pair of brass rings called gimbals, in such a manner that the bowl will remain horizontal no matter how violently the ship may pitch and roll.

The card, which is circular, is divided into degrees and thirty-two equal parts called the *points* of the compass. The magnetised needles, of which there are generally from two to four, are fastened to the bottom of the card. In the center of the card is a conical socket poised on an upright pin fixed into the bottom of the bowl so that the card, when hung on the pin, turns freely around its center. When used on a ship, the com-

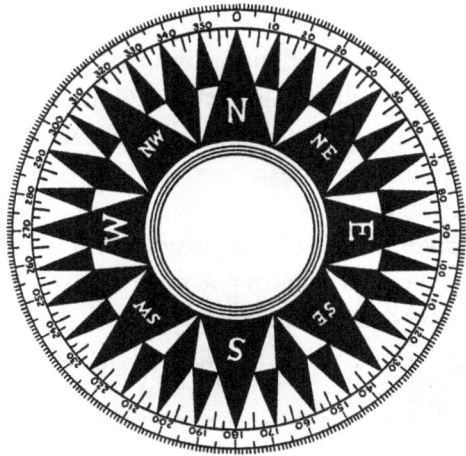

FIG. 4. — A Compass Card. The Black Diamond-Shaped Marks and Triangles Are the "Points."

pass is so placed that a black mark, called the *lubber's line,* is fixed in a position parallel to the vessel's keel. The point on the compass card which is directly against this line indicates the direction of the ship.

The compass was the first practical use found for magnetism. It was an important use because not only was it an invaluable aid to ordinary navigation but it made possible the exploration of the earth's surface.

EXPERIMENTS WITH MAGNETISM

Magnetism is very closely related to electricity. The laws which govern the action of magnets play an important part in the construction of almost all electrical apparatus. It is by means of magnetism that electricity is generated in our power houses. Dynamos, motors, telegraph instruments, telephones, radio apparatus, voltmeters, ammeters, and many other electrical devices depend upon magnetism for their operation.

Artificial Magnets are made of steel. The first artificial magnets were made by rubbing a piece of steel and a *natural* magnet or lodestone together. But artificial magnets or *permanent* magnets, as they are more commonly called, are more powerful when made with the aid of an electric current. The method of making a permanent magnet with electricity is explained in a later chapter. The process of giving a piece of steel the properties of a magnet is called magnetization.

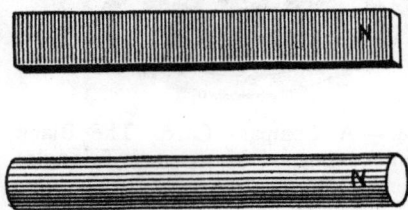

The artificial magnets used in the construction of electrical instruments are made of special steels called magnet steels. Magnets made from magnet steels are more powerful and retain their magnetism longer than magnets made from other varieties of steel.

FIG. 5. — Small Bar Magnets.

Permanent magnets are made in different shapes but the two most common ones are the Bar and Horseshoe types.

In order to secure a very powerful magnet, a number of similarly shaped pieces of steel are separately magnetised and then clamped together. A magnet made in this manner is called a *compound* magnet and may have either the bar or horseshoe shape. The magnets used on an automobile are usually compound.

Toy Magnets. Small horseshoe magnets can be purchased at toy-stores or at a dime store. They can be used to perform a number of interesting and instructive experiments. A practical knowledge of magnetism will enable you to understand the action of many electrical devices.

Toy horseshoe magnets are provided with a piece of soft iron called an armature or "keeper." The keeper is laid across the poles of the magnet when the latter is not in use. It preserves the magnetism. Without it, the magnet would in time become weak.

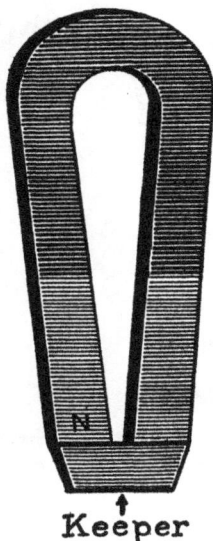

Keeper

Fig. 6. — A Toy Horseshoe Magnet.

A permanent magnet from a discarded magneto, radio speaker, or telephone receiver is useful for experimenting.

A permanent magnet may be used to produce other permanent magnets. Rub a large darning needle across one of the poles of a horseshoe magnet, stroking the needle from end to end and always in the same direction. Then dip the needle into some iron filings. The filings will cling to the needle in tufts *at the ends.* The needle has become a permanent magnet.

If you dip a horseshoe or bar magnet into iron filings, it will be noticed that the filings cling to the magnet in irregular tufts *at the ends.* with few if any near the center.

This shows that the greatest attractive power of a magnet exists at its ends. The places where the attractive power is thus concentrated are called the *poles*.

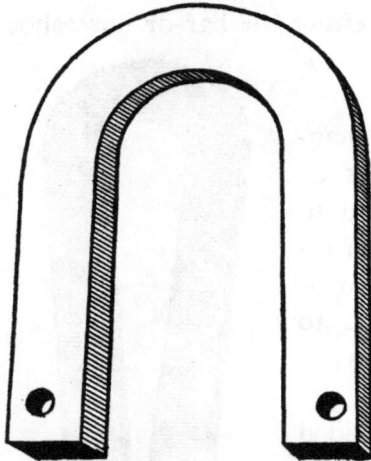

FIG. 7. — A Permanent Magnet from a Radio Loud-Speaker.

Another experiment which shows how the attractive power concentrates at the ends of a magnet can be performed with a steel knitting-needle. Magnetise the needle by stroking it across the poles of another permanent magnet. If it is dipped into iron filings, the filings will cling to the end. Then if the knitting-needle is broken in two at the center where it does not show any magnetism, each half will prove to be a magnet with a pole at each end. Iron filings will cling to the ends of each piece. If you repeat the process and break each piece again, you will find that every time a magnet is broken, new poles are formed at the break.

The strange, unseen force which can exert itself across space between magnets and pieces of iron and steel, and which we have been calling magnetism so far in these pages, is more scientifically spoken of as magnetic force.

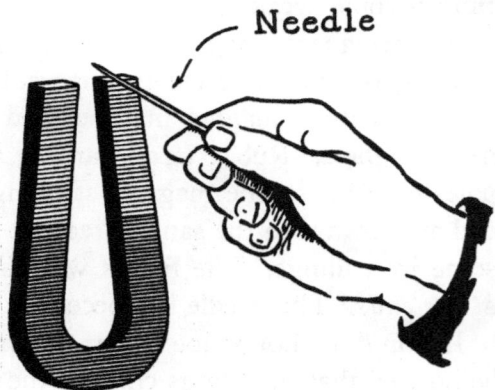

Needle

FIG. 8. — Magnetizing a Sewing Needle.

Magnetic Force. One of the important laws of magnetism is that the pull exerted by a magnet upon a piece of iron is not the same at all distances. The pull is much stronger when the

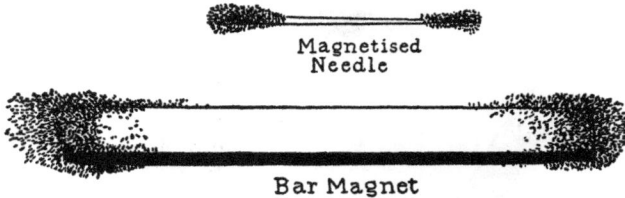

Magnetised
Needle

Bar Magnet

FIG. 9. — A Magnetized Sewing Needle and a Bar Magnet Which Have Been Dipped in Iron Filings. Notice how the Filings Cling in Tufts at the Ends.

iron is near the magnet and weaker when it is farther away.

You can prove this if you place some small carpet-tacks on a piece of paper and hold a magnet above them. Gradually lower one pole of the magnet until the tacks jump up to meet it. Notice how far the magnet is from the tacks when they are lifted.

Then try the same experiment, using a large nail in place

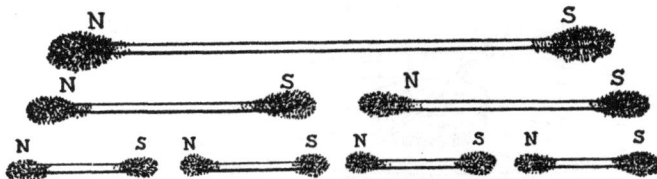

FIG. 10. — An Experiment Showing That When a Magnet is Broken, New Poles Form at the Ends of the Pieces.

of the tacks. The nail is heavier than the tacks and will require more force to lift. The pole of the magnet must be brought much closer to the heavy nail than to the light tacks before it is lifted. This shows that the magnetic force exerted by a magnet is strongest close to it.

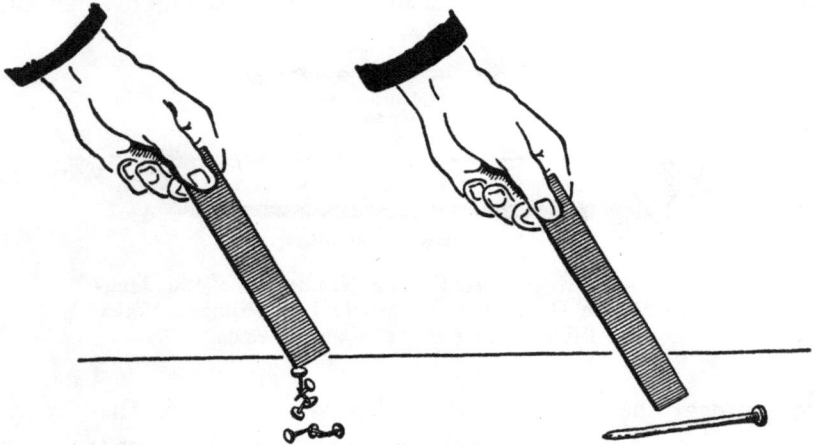

FIG. 11. — An Experiment to Illustrate the Lifting Power of a Bar Magnet. The Magnet Must be Brought Closer to the Nail Than the Tacks Because it Requires a Stronger Force to Lift the Heavier Nail.

Magnetic Poles. The poles of a magnet look exactly alike and act exactly alike except in one respect. You can detect the difference if you magnetise a darning-needle and lay it on a flat cork floating in a glass or porcelain vessel partly filled with water. It will be noticed that the needle always comes to rest lying in a north and south position. The same end always points toward the north. All magnets which are free to

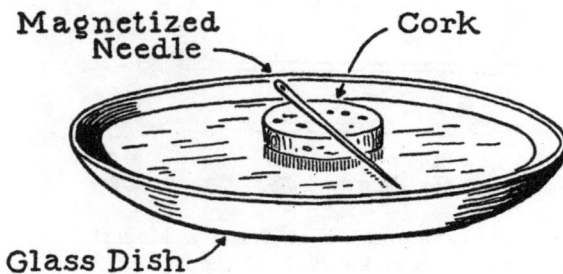

FIG. 12. — A Simple Compass. A Magnetized Needle Resting on a Floating Cork.

move without much friction, such as a magnetised needle on a cork floating in water, tend to swing around and come to rest in a north and south position. Since the same pole always points to the north, there must be some difference between the poles.

The pole which tends to turn towards the north is called the *north-seeking pole*. These names are usually abbreviated to north and south poles. The north pole of a magnet is usually marked by a straight line or the letter N stamped in the metal.

A magnetised needle on a cork floating in a basin of water is a simple compass like the gnomon used by medieval mariners.

How to Make Magnetic Compasses. A magnetised darning-needle suspended from a fine silk fiber or thread will also act as a compass and swing around until it rests in a north and south position.

A more sensitive compass may be built with two magnetised sewing needles, two pieces of paper, a cork, and a pin. The paper should be cut according to the pattern shown in Fig. 14. The two magnetised needles are stuck through the sides of the paper piece marked B. The north poles of the needles should be at the same end of B. The paper support rests on a pin stuck through a flat cork.

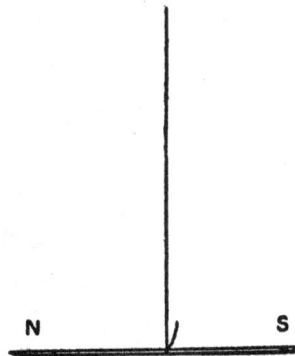

FIG. 13. — A Compass Made by Suspending a Magnetized Sewing Needle with a Thread.

A simple compass may also be made from a small piece of watch-spring or clock-spring about one and one-half or two inches long. The piece of spring is used as the compass needle. The center of the spring must be softened or *annealed* by holding it in the flame of an alcohol lamp until it is red hot. Then allow it to cool slowly. Lay the needle on a piece of soft metal, such as copper, brass, or lead, and dent it deeply at its exact center with a center-punch. Magnetise the needle and balance it on the point of a pin stuck through a flat cork.

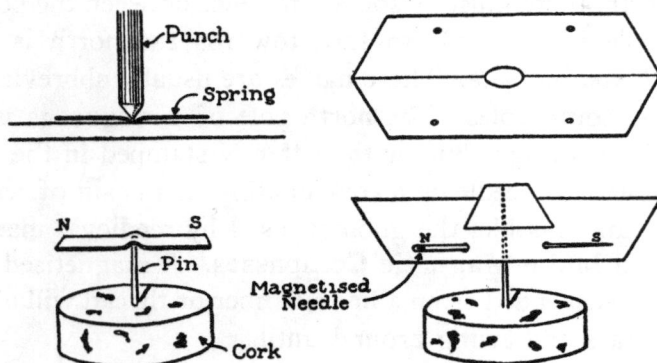

FIG. 14. — Two Methods of Making a Simple Compass.

What Are Magnetic Substances? Magnetic substances are those which are attracted by a magnet. Experiment with several different materials such as paper, wood, brass, iron, copper, zinc, rubber, steel, chalk, etc.

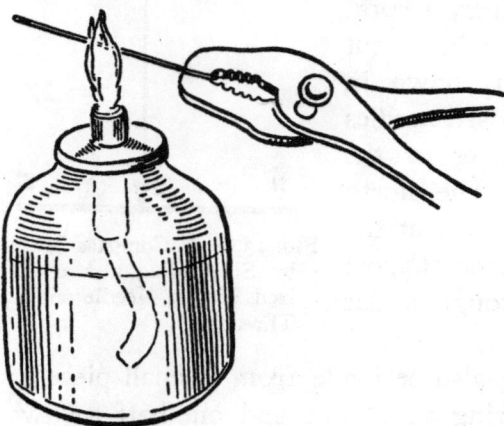

FIG. 15. — Heating a Magnetized Needle Destroys Its Magnetism.

You will find that only iron and steel are capable of being attracted by your magnet. Magnets of ordinary strength attract but very few substances to a noticeable extent. Cobalt and nickel are attracted slightly.

With the exception of some special alloys or mixtures of

metals which have recently been developed for making magnets, *hardened* steel is the only material which will retain any appreciable amount of magnetism.

You can make a magnet of a knife-blade, screwdriver, or file because these articles are made of *hardened* steel.

Heat causes a magnet to lose its magnetism.

If you heat a magnetised needle until it is red hot, taking care to heat every part of it, the needle will lose its magnetism, and iron filings will no longer cling to it. Heating the needle red hot and allowing it to cool changes the hardened steel needle into a soft steel needle. Soft steel does not retain magnetism.

Does Magnetic Force Pass through Substances? The attractive force of a magnet will make itself felt through paper,

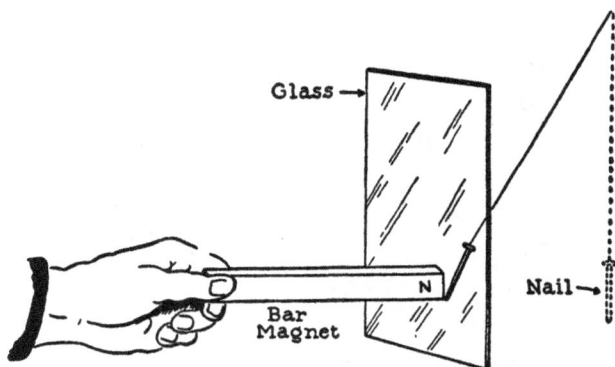

FIG. 16. — An Experiment Which Shows the Attraction of an Iron Nail Through Glass.

glass, wood, brass, and practically all other substances except iron or steel. The attraction is reduced or entirely checked through an iron or steel plate, because the iron or steel absorbs the magnetic effect itself and prevents the force from passing through and reaching the nail.

If you are interested in electricity, you will no doubt often read or hear the word *induction*. It means influence. A magnet by induction is a magnet because of the influence or proximity of another magnet. Magnetic induction, electrostatic induction and electromagnetic induction are familiar words to an electrical engineer. Electrostatic and electromagnetic induction are explained later in this book.

Magnetic Induction. Here is an experiment which will explain magnetic induction. A number of carpet tacks may be supported from one pole of a magnet in the form of a chain. Each tack supports the one below it. Each individual tack becomes a temporary magnet by *induction*. If the tack next to the magnet is grasped with a pair of tweezers and the magnet suddenly pulled away, the tacks will at once lose their magnetism and fall apart. The magnet is no longer there to exert its influence or inductive action.

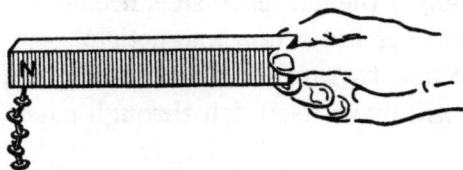

FIG. 17. — A Magnetic Chain.

A magnet will only support a certain number of tacks in the form of a chain. But if a second magnet is placed beneath the chain, so that its south pole is under the north pole of the original magnet, the chain may be lengthened by the addition of several more tacks. There is a simple reason for this. The temporary magnetism in the tacks is further increased by in-

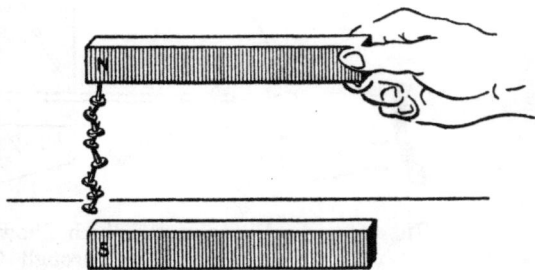

FIG. 18. — The Strength of the Chain is Increased by the Induction of a Second Magnet.

duction or the influence of the second magnet.

Magnets Will Repel as well as attract each other depending on which poles are nearest. Here is the way you can prove it.

Magnetize a sewing-needle and suspend it with a thread which passes through the eye. Bring the north pole of a bar or horse-shoe magnet near the lower end of the needle. If the lower

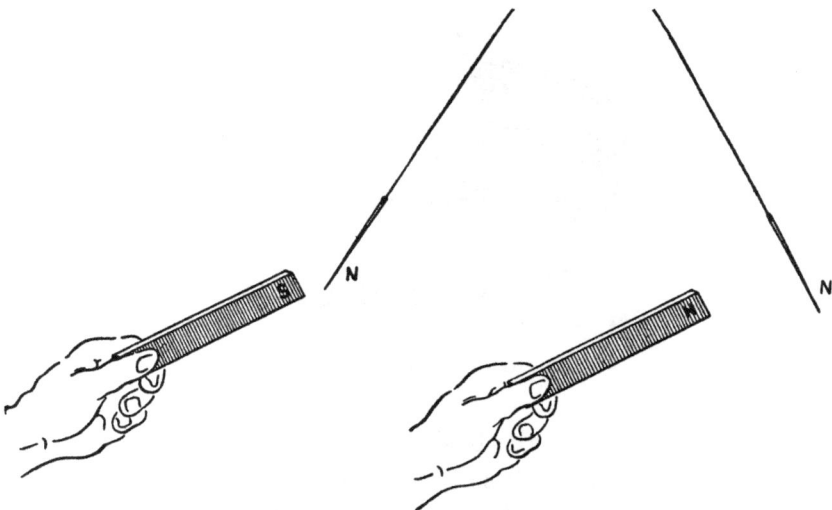

FIG. 19. — An Experiment Showing that Similar Poles Repel Each Other and Unlike Poles Attract.

end of the needle happens to be a *south* pole, it will be attracted by the *north* pole of the bar magnet. If, on the other hand, it is a north pole, it will be repelled and you cannot touch it with the north pole of the magnet in your hand unless you catch it and hold it.

If there is such a thing as a "general" law of magnetism it is that :

Like poles repel each other and unlike poles attract each other.

An interesting method of illustrating this important law of magnetism is by making a small boat from cigar-box wood and laying a small bar magnet on it. Place the bar magnet so that the north pole points towards the bow of the boat.

Float the boat in an *aluminum* basin. Bring the south pole of a bar magnet near the stern of the boat and it will sail away

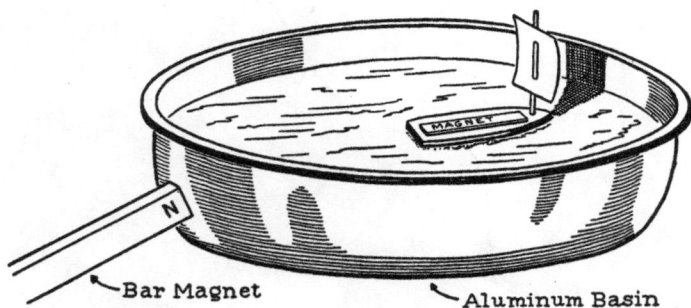

FIG. 20. — A Magnetic Boat.

to the opposite side of the basin. Present the north pole of the magnet and the boat will sail back again.

If the south pole of the magnet is presented to the bow of the boat, the little ship will follow the magnet around the basin.

The fact that similar poles of magnets repel each other may be illustrated by a number of magnetised sewing-needles fixed in small corks so that they will float in a basin of water with their points down.

The needles and corks will arrange themselves in symmetrical groups, according to their number. The various patterns which will be formed are shown in one of the illustrations. The upper ends of the needles should all have the same polarity; that is, be either all north or all south.

A bar magnet thrust among the floating needles will attract

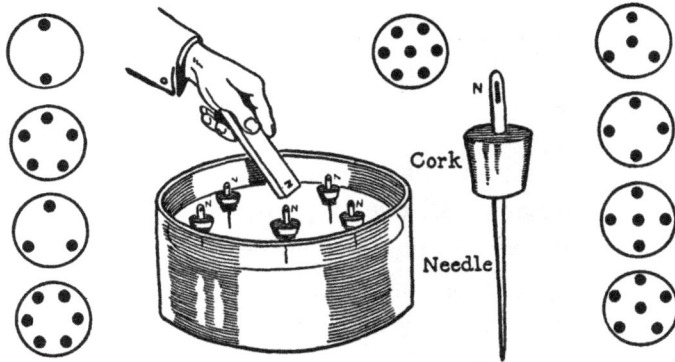

Fig. 21. — The Repulsion of Similar Magnetic Poles, Shown by Floating Needles.

or repel them depending on the polarity of the lower end of the magnet.

What Is in the Space around a Magnet? You can answer this question with a simple experiment. Magnetism "flows" in

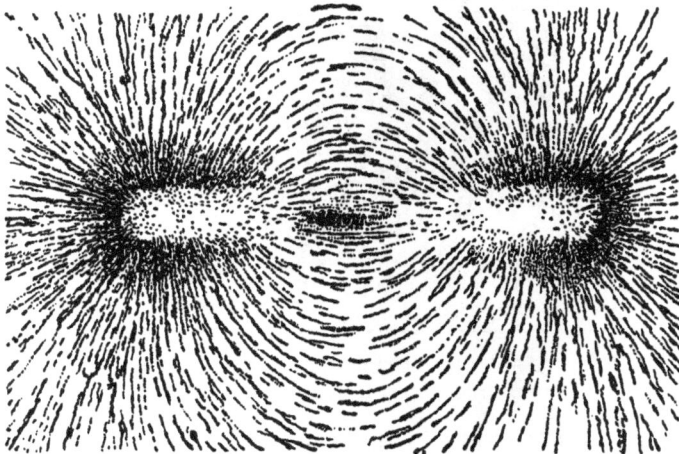

Fig. 22. — A Magnetic "Phantom," Showing the Field of Force about a Bar Magnet.

lines called *lines of magnetic force*. The space around a magnet through which these lines pass is called the *field of force* and the path through which they "flow" is called the *magnetic circuit*.

The magnetic circuit may be seen by placing a sheet of stiff paper, or a pane of glass, over a permanent magnet and sprin-

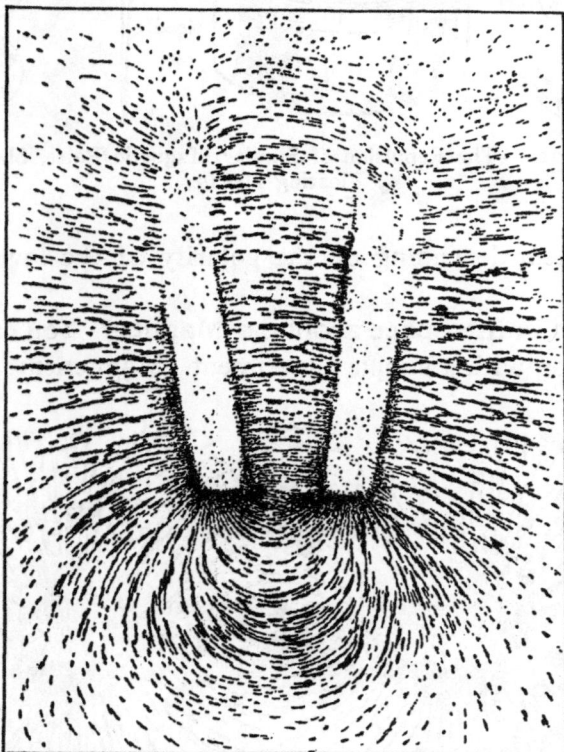

FIG. 23. — Magnetic Phantom Showing the Lines of Force about a Horseshoe Magnet.

kling iron filings over the paper. When the paper is tapped slightly, the filings will form into lines showing the magnetic circuit.

The lines of filings diverge from one pole and curve around to the opposite pole.

In thus outlining the paths of the lines of force, you have made what is called a magnetic *phantom*.

Fig. 22 shows the magnetic phantom formed about a bar magnet. The phantoms of two bar magnets, one with north

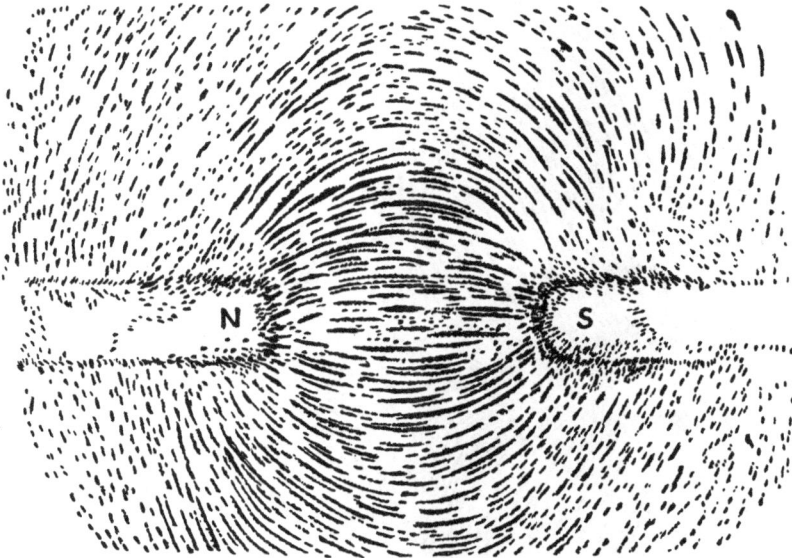

FIG. 24. — Phantom Showing the Paths of the Lines of Magnetic Force Between Unlike Poles.

and south poles opposite and one with two north poles opposite, are shown in Figs. 24 and 25. These plainly show the attractive action between unlike poles and the repelling action of like poles. The lines between unlike poles pass in regular curves but the lines of force between like poles form, roughly, a diamond-shaped space free of magnetic lines.

The Earth Is a Great Magnet. The action of the earth on a compass needle is exactly the same as that of a permanent magnet because the earth itself is a magnet with lines of force passing in a north and south direction.

The magnetic pole of the earth towards which the compass needle points is not situated at the same place as the geographic pole. The compass needle therefore does not point exactly toward the true North from all spots of the earth. The direction assumed by a compass needle when it comes to rest in a north and south position is called the *magnetic meridian.*

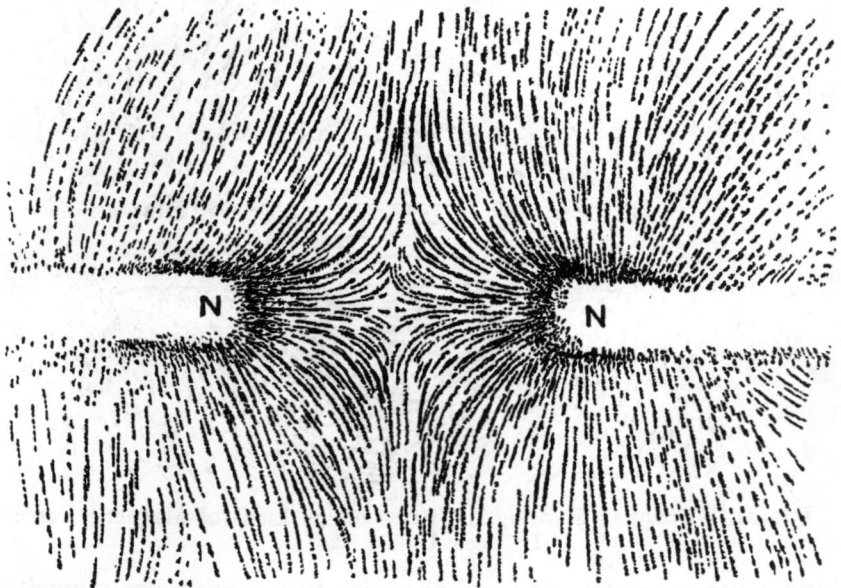

FIG. 25. — Phantom Showing the Paths of the Lines of Magnetic Force Between Similar Poles.

Magnetic Dip. If a long needle is carefully balanced so as to be perfectly horizontal when supported at its center, and then magnetised, it will lose its balance. The north-seeking end will point slighly downward in those parts of the earth which are north of the equator. This is due to the fact that the earth is round.

At the north magnetic pole, the needle will point directly downward.

A magnetic needle, mounted so that it can swing up and down, instead of from side to side like a compass needle, is called a *dipping* needle.

An Experimental Dipping Needle may be made by thrusting a steel knitting-needle through a cork before it has been magnetised. A second needle is thrust through the cork at right angles to the first. The arrangement is carefully balanced by shifting the position of the first needle until it will remain horizontal when resting on the edge of two glass tumblers or wine glasses.

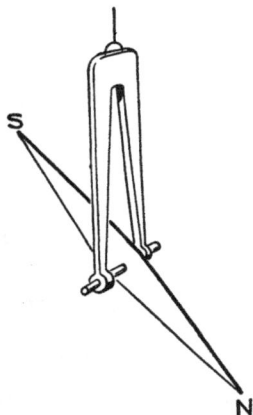

FIG. 26. — A Dipping Needle.

Then magnetize the first needle by stroking it with a magnet. When it is again allowed to come to rest on the edges of the glasses, it will no longer balance, but dip downward.

How to Preserve the Strength of a Permanent Magnet

FIG. 27. — An Experimental Dipping Needle.

A sharp blow or a fall will weaken a permanent magnet. All permanent magnets will lose part of their magnetism if heated and will lose all of it if they become red hot.

If you wish to preserve the strength of your magnets, handle them gently. When a magnet is not in use, it should be laid on a piece of iron.

Chapter II

STATIC ELECTRICITY

How We First Became Acquainted with Electricity

If you rub a glass rod or a stick of sealing wax with a piece of dry, warm flannel, it will be found to have acquired a property which it did not formerly possess: namely, the power

Fig. 28. — An Electrified Stick of Sealing-Wax Will Attract Small Bits of Paper.

of attracting to itself such light bodies as dust or bits of thread and paper.

Hold a stick of sealing wax which has been rubbed with a piece of flannel over some small bits of paper and watch them jump up to meet it just as if the sealing wax were a magnet attracting small pieces of iron instead of paper.

The agency at work which gives the sealing wax this mysterious power is called *electricity*. The name came from the Greek word "Elektron," which means amber. Amber was the first substance found to possess the property when rubbed.

It is always interesting to know the history of things. When we look into the history of amber, we find that its use began in very early times. Amber beads, at least 2000 years old, have been found in the royal tombs at Mycenae and at various places in Sardinia.

Amber was used by the ancient world as a jewel, and for decoration.

The ancient Syrian woman sometimes used distaffs made of amber for spinning. As the spindle whirled around it often rubbed against the spinners' garments and thus became *electrified*. On nearing the ground it drew to itself the dust or bits of chaff and leaves lying there, or sometimes attracted the fringe of the clothing.

The spinner easily noticed this because the bits of chaff which were thus attracted would become entangled in her thread unless she was careful. The amber spindle was, therefore, called the *harpaga* or "clutcher," for it seemed to seize such light bodies as if it had invisible talons.

This was probably the first intelligent observation of an electrical effect.

Of course there had always been lightning flashing about in the sky. We know that lightning is a flash of electricity.

Benjamin Franklin was the first to prove that lightning discharges are electrical. The story of his experiment is very interesting.

Believing lightning to be electricity, Franklin planned to

draw it out of the clouds with a wet kite string. He made a kite of two light strips of cedar wood, covered with a large silk handkerchief. To the top of the upright stick of the kite was fastened a sharp wire. The twine attached to the kite was the usual kind, but was provided with a short piece of silk ribbon and a key. The purpose of the silk ribbon was the possible protection which it, as an insulator, might afford in preventing the lightning from shocking the kite flyer. The key was secured to the junction of the silk ribbon and the twine to serve as a convenient conductor from which to draw sparks—if they came. He did not have to wait long for a thunderstorm.

When he saw it gathering, Franklin went out into a field with his son, William, then a young man twenty-two years of age. Great clouds were rolling up from the horizon and the gusts of wind were becoming fitful and strong.

The kite rose steadily into the air, swooping this way and that as the breeze caught it. The thunder muttered nearer and the rain began to patter on the ground as the kite flew higher.

The rain soon began falling heavily and Franklin and his son were compelled to take refuge under a near-by shed. The kite, wet with rain, was sailing sluggishly.

Franklin watched the silk ribbon and the key. There was not a sign. Had he failed? Suddenly the loose fibers of the twine erected themselves. The moment had come. Without a tremor he advanced his knuckle to the key. And between his knuckle and the key passed a spark! then another and another. They were the same kind of little sparks that he had made hundreds of times with his static machines.

And then as the storm abated and the clouds cleared away, Franklin and son William wound in the kite string. The experiment was successful. He could tell the world that he had proven lightning to be an electric spark. No longer would it be a mystery.

Knowing what it was, practical means of protecting buildings against lightning could be devised. So Franklin invented the lightning rod.

The Cause of Lightning is the accumulation of electric charges in the atmosphere, the electricity gathering on the surface of the tiny particles of water in the cloud. These charges grow stronger as the particles of water join together and become larger. As the countless multitude of drops grows larger the pressure or *potential* of their electric charge is increased and the cloud becomes more heavily charged.

The charge also grows stronger through the inductive action, or influence, of the electricity residing on a neighboring cloud or on the earth below. There is the phenomenon called electrostatic induction which was mentioned in Chapter I.

The charge on a cloud continually tries to burst across the intervening air. As soon as the charge grows strong enough a vivid flash of lightning, which may be from one to ten miles long, occurs. The heated air in the path of the enormous electric spark expands with great force; but immediately other air rushes in to fill the partial vacuum, thus producing the terrifying sounds called thunder.

What Is Static Electricity? In the eighteenth century, electricity was believed to be a fiery discharge. Later, it was thought to be some sort of invisible fluid. There has been reason to believe that it consists of the motion or vibration of tiny particles. At the time it was suggested, there had been evidence to support each of the many explanations of the nature of electricity. The theory which modern scientists hold is too complicated to attempt to explain in a boys' book.

The electricity of the lightning and the electricity generated by rubbing glass rods, sealing wax, etc., is called *static* electricity from the Greek word meaning "standing" because it is generally at rest or standing still on the surface of things. Static electricity is not of much practical use.

The electricity which furnishes light, operates telephones, motors, etc., is not static electricity produced by magnetism or the chemical action of batteries, and is explained in later chapters.

It was early discovered that electricity would travel through some mediums but not through others. These were termed respectively *conductors* and *non-conductors* or insulators. Metals such as silver, copper, and gold, and other substances such as charcoal and water, are conductors. Glass, silk, wool, oils, and wax are non-conductors or insulators while many other substances such as wood and marble are partial conductors.

EXPERIMENTS WITH STATIC ELECTRICITY

A simple method of generating static electricity is by shuffling or sliding the feet over the carpet. The body will then become *charged,* and if the knuckles are presented to some metallic object, such as a radiator, a snapping little spark will jump between. The electricity is produced by the friction of the shoes sliding over the woolen carpet.

Paper

FIG. 29. — A Piece of Dry Paper May Be Electrified by Rubbing with the Hand.

Static electricity is easily produced by friction.

Warm a piece of writing paper, then lay it on a wooden table and rub it briskly with the hand. It soon will stick to the table and will not slide along as it did at first. If one corner is raised slightly it will tend to jump right back. If the paper is lifted off the table, it will try to cling to the hands and clothing. If held near the face it will produce a tickling sensation. All these things happen because the paper is electrified. It is drawn to

FIG. 30. — A Surprise for the Cat.

the other objects because they are neutral; that is, do not possess an electrical charge.

All experiments with static electricity perform better in the winter time, when it is cool and clear, than in the summer. The reason is that the air is drier in winter than in summer. Summer air contains considerable moisture and water vapor. Water vapor is a partial conductor of electricity, and the surrounding air will therefore conduct the static electricity away from your apparatus almost as fast as it can be produced in the summer.

Some day during the winter, when it is cool and clear, stroke the cat rapidly with the hand. The fur will stand up towards the hand and a faint crackling noise will be heard. The crackling is caused by small sparks passing between the cat and the hand. If the experiment is performed in a dark room, the sparks may be plainly seen. If you present your knuckle to the cat's nose a spark will jump and somewhat surprise the cat.

If the day is brisk and cool, so that everything outside is frozen and dry, comb your hair briskly with a hard rubber comb and notice what happens. Your hair will stand up instead of lying down flat, and the faint crackling noise, showing that sparking is taking place as the comb passes through the hair, will be plainly heard. The electricity is produced by the friction between the hair and the comb.

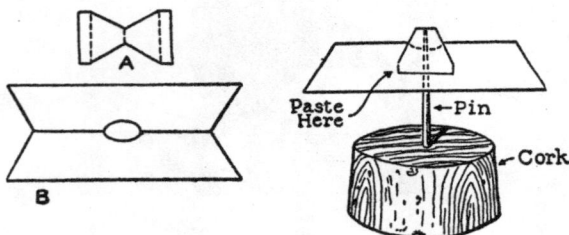

FIG. 31.— A Paper Electroscope.

Static electricity for experimenting may be produced by friction between a number of common substances. A hard rubber comb, a glass rod, a stick of sealing-wax may be easily electrified by rubbing briskly with a piece of dry, warm wool, flannel, fur, or silk.

Electroscopes are devices for detecting the presence of static electricity.

A very simple form of electroscope may be made in much the same manner as the paper compass described in the last chapter. It may be cut out of stiff writing-paper and mounted on a pin stuck through a cork. If an electrified rod is held near the electroscope, it may be made to whirl around in the same manner as a compass needle when a bar magnet is brought close.

The Pith-Ball Electroscope is an instrument in which a small ball or cork or elder pith is hung by a fine silk thread from an insulated support. An experimental electroscope may be made from a glass bottle having a piece of stiff wire thrust into the cork to support the pith ball. The glass bottle serves as an insulator.

When an electrified comb, stick of sealing wax, or glass rod is presented to the pith ball, it will fly out towards the comb. If the pith ball be permitted to touch the comb, the latter will

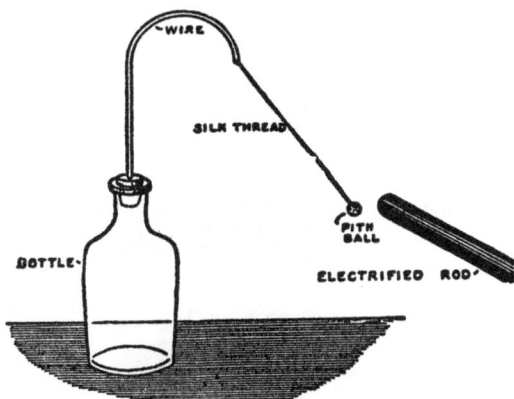

FIG. 32. — The Pith-Ball Electroscope.

transfer some of its electricity to the ball. Almost immediately the pith ball will fly away from the comb, and no matter how near the comb is brought, it will refuse to be touched again until it has lost its charge.

This action is much the same as that of the magnetized needle suspended from a thread when the similar pole of a magnet is presented to one of its poles.

When the comb is first presented to the pith ball, the latter is neutral; that is, carries no charge. When the comb has touched the ball, however, some of the electricity from the rod passes to the ball. After that happens they repel each other.

The reason that the comb repels the pith ball after they have touched is that they are *similarly* charged and *similarly charged bodies repel each other.*

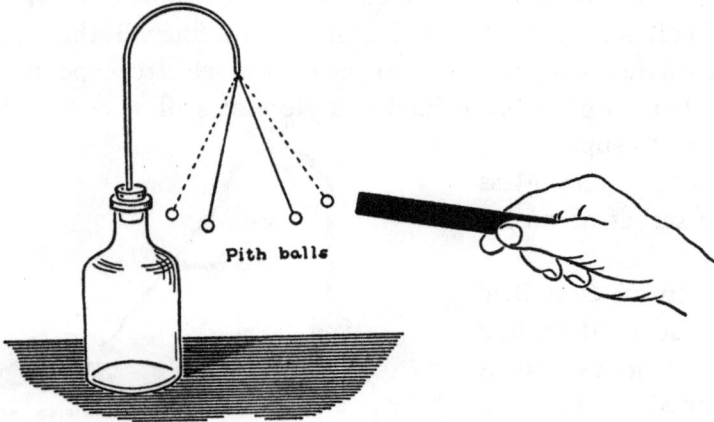

FIG. 33. — A Double Pith-Ball Electroscope.

If you are a good observer, you might have noticed when experimenting with an electrified rod and the small bits of paper, that some of the bits of paper were first attracted and flew upwards to the rod, but having once touched it, were quickly repelled.

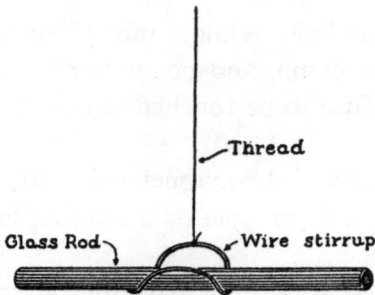

FIG. 34. — Method of Suspending an Electrified Rod in a Wire Stirrup.

The repulsion between two similarly electrified bodies may be shown by a *double electroscope.*

A double electroscope is made by hanging two pith balls on two silk threads from the same support.

Electrify a stick of sealing wax, a comb, or a glass rod, and touch it to the pith balls. They will immediately fly apart because they are electrified with the same kind of electricity.

There are two kinds of static electricity. Rub a glass rod with a piece of silk and suspend the rod in a wire stirrup hung by a thread as shown in the illustration. Electrify a second

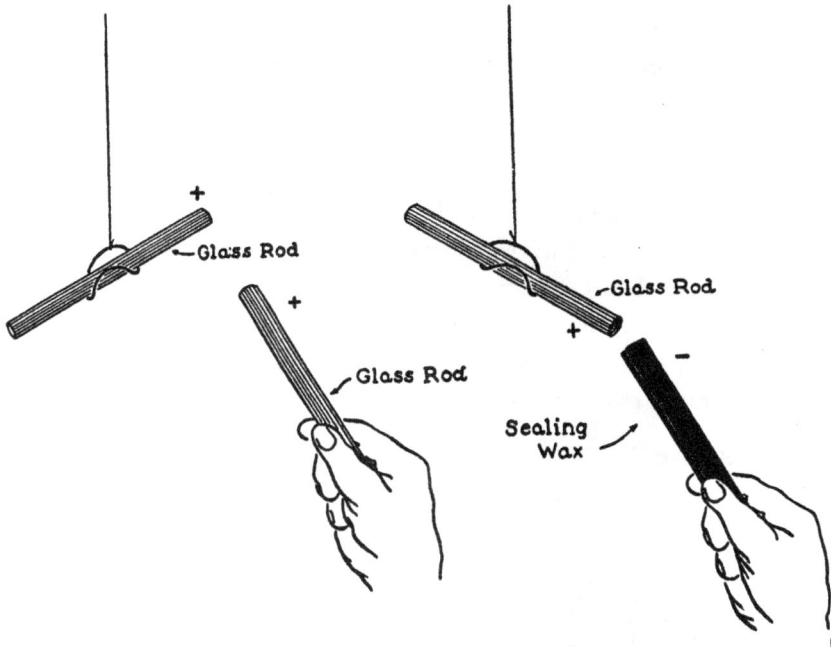

FIG. 35. — An Experiment Showing that Similarly Electrified Bodies Repel Each Other. Dissimilarly Electrified Ones Attract Each Other.

rod also and bring it near one end of the rod suspended in the stirrup. The suspended rod will be repelled and will swing away from the one held in the hand. Now rub a stick of *sealing wax* with a piece of flannel until the sealing-wax is electrified. Then bring the stick of sealing-wax near the end of the suspended glass rod. The rod will be *attracted* to the sealing-wax.

If you experiment further, you will find that two sticks of sealing-wax will repel each other.

These experiments indicate that there are two kinds of electrification: *positive* electricity, as developed by rubbing glass with silk, and *negative* electricity, developed by rubbing sealing-wax with flannel.

It can be shown that the materials used as rubbers, the silk and flannel, are also electrified, the silk being *negative* and the flannel *positive*.

The same law that applies to magnetism also holds true in the case of static electricity, and similarly electrified bodies will *repel* each other and *dissimilar* ones *attract*.

The Gold Leaf Electroscope is one of the most sensitive means of detecting feeble charges of static electricity.

It is a simple instrument and one is easily made in a short time. Two narrow strips of the thinnest tissue paper, which have been rubbed with aluminum powder, or, better still, two strips of gold leaf, are hung from a support in a wide-mouthed glass bottle which serves at once to insulate and protect the strips from draughts of air.

The mouth of the jar is closed by a plug of paraffin wax through the center of which passes a small glass tube. The lower end of the wire is bent at right angles to furnish support for the strips of gold leaf. A round sheet metal disk about the size of a quarter is soldered to the upper end of the rod.

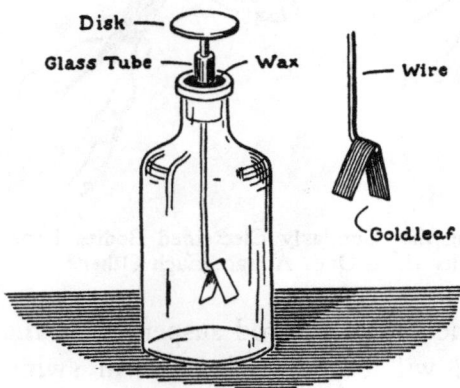

FIG. 36. — A Gold-Leaf Electroscope.

If an electrified stick of sealing-wax or a glass rod is presented to the disk of the electroscope, the gold leaf strips will repel each other very strongly. If the instrument is sensitive and thoroughly dry, the strips should begin to diverge as the rod approaches the disk before it touches. It is possible to make a gold leaf electroscope so sensitive that chips formed by sharpening a pencil will cause the strips to diverge.

The Electrophorus is an instrument devised by Volta in 1775 for the purpose of obtaining static electricity.

It is easily constructed and will furnish a source of electricity for many interesting experiments. An electrophorus consists of two parts: a round cake of resinous material cast in a metal disk or pan, and a round metal disk which is provided with an insulated handle.

To make an electrophorus, first procure an old cake or pie tin, and fill it with bits of resin or sealing-wax. Place the pan in a warm spot on the stove where the resin will melt, taking care not to overheat it or it will possibly take fire. As the resin melts, add more until the pan is nearly full. When all is melted, remove from the fire and set it away where it may cool and harden in the pan without being disturbed.

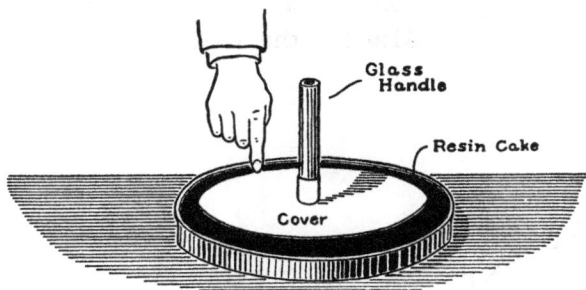

FIG. 37. — Volta's Electrophorus.

Cut a circular disk out of sheet tin, zinc, or copper, making the diameter about two inches less than that of the pie pan. Solder a small cylinder of tin or sheet brass to the center of the disk to aid in supporting the handle. The latter is a piece of glass tubing about one-half inch in diameter and four or five inches long, placed in the center of the cylinder and secured in place with molten sealing-wax.

In order to use the electrophorus the resinous cake must first be rubbed with a piece of warm flannel, woolen cloth or preferably fur. Then place the disk on the cake, holding the insulating handle with the right hand. Touch the cover or disk momentarily with the fore-finger of the left hand. After the finger has been removed, raise the disk from the cake by picking it up with the glass insulating handle. The disk will now be found to be heavily charged with positive electricity, and if the knuckles are presented to the edge, a spark will jump out to meet them.

The cover may then be replaced, touched, and once more raised. It will yield any number of sparks, the resinous cake only needing to be recharged by rubbing once in a while.

An Electric Frog-Pond may be experimented with by cutting out some small tissue paper frogs of the size shown in the illustration. Moisten the frogs slightly by breathing on them

Tissue paper frog

FIG. 38. — An Electric Frog-Pond.

and then lay them on the cover of the electrophorus. Touch the electrophorus with the finger and then raise it with the insulating handle. If the "frogs" are not too wet, they will jump off the cover as soon as it is raised.

This is another example of the electrical law at work; that similarly charged bodies repel each other.

Chapter III

STATIC ELECTRIC MACHINES

A Cylinder Electric Machine

An electrophorus will furnish sufficient static electricity for many interesting experiments but when larger supplies of electricity are necessary a static electric machine is necessary.

Static electric machines are of two kinds: *Influence Machines,* and *Frictional Machines.* Electrostatic induction plays a part in producing electricity in an influence machine. In a frictional machine, as the name would indicate, electricity is produced by friction.

The earliest form of electric machine consisted of a ball of sulphur fixed upon a spindle which could be rotated by means of a crank. When a person standing on an insulating cake of resin pressed his hands against the revolving sulphur globe, sparks could be drawn from his body. Later on, the machine was improved. A leather pad was substituted for the hands, and a glass globe or cylinder for the ball of sulphur.

Essentially, a frictional electric machine consists of two parts, one for producing the electricity by the friction of two surfaces rubbing against each other, and the other an arrangement for collecting the electricity thus formed.

In the machine which we will build, the electricity is produced by the friction between leather and glass.

A leather cushion, stuffed with horsehair and covered with

48

a powdered amalgam of zinc or tin, presses against one side of a revolving glass bottle. A "prime" conductor in the shape of an elongated cylinder presents a row of fine metal points, like the teeth of a rake, to the opposite side. A flap of silk attached to the leather cushion covers the upper half of the cylinder.

When the handle of the machine is turned, the friction between the leather cushion and the revolving glass bottle generates a supply of positive electricity on the glass, which is collected by the row of sharp points on the prime conductor.

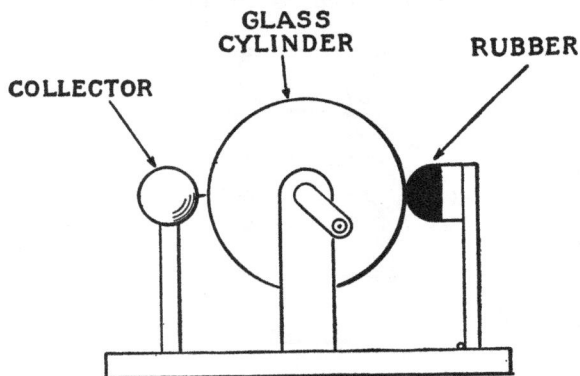

FIG. 39. — Front View of a Cylinder Electric Machine.

The first thing required in the construction of an electric machine is a large *white* glass bottle having a capacity of at least two quarts and preferably one gallon.

Select a smooth bottle which has no lettering embossed upon it, and set it on a piece of white paper. Carefully trace a line around the circumference of the bottle so that the circle thus formed on the paper is exactly the same size as the bottom of the bottle.

Lay a carpenter's square on the circle so that the point C just touches the circumference. Draw a line from A to B where the sides of the square cut the circumference. The point in the middle of this line is the center of the circle. Place the paper on the bottom of the bottle so that the circle coincides with the circumference exactly, and mark the center of the bottle. You can mark on glass with the sharp edge of a small broken file.

It is important to do this work with the utmost care because the mark is to be used to locate a hole in the bottom of the bottle. If the hole is not in the center, the bottle will wobble.

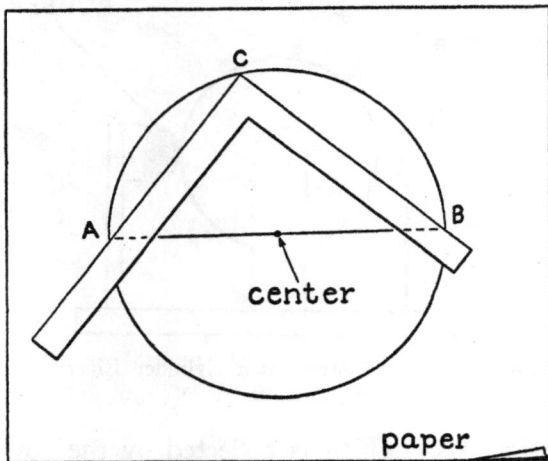

FIG. 40. — Method of Finding the Center of the Bottom of the Bottle.

It is not easy to drill a hole in the bottom of the bottle. It is done with a copper or brass tube, slightly smaller in diameter than the desired hole. The hole should be about three-eighths of an inch in diameter. The tube serves as a drill. It is kept smeared with a mixture of turpentine and emery or Carborundum powder. You may use a coarse valve-grinding compound such as auto mechanics use for grinding engine valves. The tube is placed in a hand-drill or drill-press and is guided to the center of the bottle by a block of wood. The wooden block is drilled with a hole which will just slip over the tube. The center of the hole is directly over the center of the bottle. The wooden block is cemented firmly to the bottom of the bottle with hot sealing-wax.

It is the grinding compound that does the actual cutting of the glass and the tube must be pulled away from the glass occasionally so that the grinding material will work in between it and the glass.

The shops which replace broken automobile windshields and windows are usually equipped to drill holes in glass. If you can afford to pay to have a hole drilled in the bottom of a bottle, you can save yourself a lot of trouble by having the work done.

The dimensions of bottles vary so that it is quite impossible to give exact dimensions for constructing all the parts of the static machine. Those given in the text will suit the average one-gallon bottle.

The neck of the bottle is fitted with a wooden plug which slides into place like a cork. If you do not have a wood-turning lathe, it may be necessary to make this part of your machine in the school manual training shop.

The axis of the plug is drilled with a three-eighths inch hole so that it will slip over the rod which forms the shaft for the bottle. The plug is cemented in the neck of the bottle. Dip the plug in shellac and allow it to dry. Then smear the inside of the bottle neck with thick shellac and push the plug in until it is snug. Allow the bottle and plug to remain undisturbed for a day or two so that the shellac will have time to set.

It is important that the plug be fastened securely. Shellac as a cement is not so dependable as the mixture which scientists used two hundred years ago in making electrical machines. Here is the formula:

½ lb. Resin
5 oz. Beeswax
¼ oz. Plaster of Paris
¾ oz. Red Ocher

Melt these ingredients in a can over a moderately hot stove. Stir them well. Warm the neck of the bottle. Dip the end of the plug into the hot cement and press it into the neck of the bottle. When the cement cools, it will be almost impossible to pull the plug out.

The end of the plug should extend two and one-quarter inches above the top of the bottle.

A wooden base will be required to mount the machine on. Two uprights are fastened to the base to serve as bearings for the glass bottle. One bearing is drilled with a three-eighths inch hole to receive the end of the shaft. The other is notched so that the wooden plug fitted into the neck of the bottle will rest in it.

The end of the plug projecting through the bearing is squared to fit a square hole in the wooden crank. The handle of the crank is an ordinary spool having one flange cut off and mounted with a screw and washer.

The rubber is a piece of wood one inch square and from six to eight inches long. A piece of soft undressed leather is tacked on as shown in the illustration and stuffed with horsehair to make it springy. You can obtain a wad of horsehair at an upholsterer's shop. The wood is shellacked and covered with tinfoil previous to tacking on the leather. A strip of wood two inches wide and one-half inch thick is fastened to the back of the rubber. The strip should be long enough so that when the lower end rests on the base, the rubber is level with

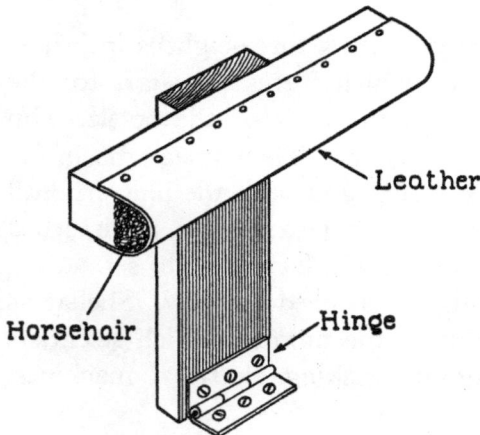

FIG. 41. — The Rubber.

the axis of the bottle. The lower end may be fastened to the base by means of a two-inch butt hinge. A heavy rubber band is arranged so that its tension will pull the rubber firmly against the bottle.

The prime conductor is formed from a piece of wooden curtain pole two inches in diameter and eight inches long. The ends are rounded with a rasp and then smoothed with sandpaper. The whole surface is then shellacked and covered with a layer of tinfoil. The tinfoil must be smooth so that there are no sharp corners or edges exposed. Static electricity will escape quickly from corners, points and edges.

The heads of a number of dressmaker's pins are cut off, and the pins forced into the side of the prime conductor with a pair of pliers. They should form a row like the teeth of a rake about three-eighths of an inch apart.

A hole is bored in the center of the under side of the prime conductor to receive the end of a piece of water-gauge glass tube. This is a strong thick-walled tube which may be purchased at a hardware store.

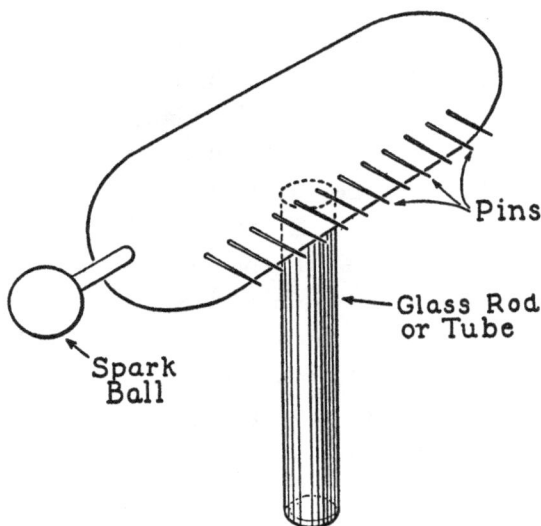

FIG. 42. — The Prime Conductor or Collector.

The lower end of the tube fits into a second hole of the same size bored into the base in such a position that when the tube is in place, the teeth on the prime conductor are on a level with the axis of the bottle, and their points about 3/32 of an inch away from the surface of the bottle. The glass tubing is necessary in order to insulate the prime conductor and prevent the escape of the electrical charge. The inside of the tube must be dry, and free from dust and dirt. You can clean it by making a swab of dry cotton rag on a wire and pushing it through the tube. The tube should be cemented in the prime conductor and also in the base.

A piece of oiled silk, eight or nine inches wide and long enough to reach half-way around the bottle, is tacked to the rubber so that it covers the upper half of the bottle to about one-half inch of the steel points. The silk is not necessary for the operation of the machine but it will increase the amount of electricity which is generated.

The machine is now complete, and ready to test. When the handle is turned rapidly so that the surface of the bottle moves *upward* against the rubber and *downward* as it passes the pin points on the prime conductor, you should be able to draw sparks from the latter with your finger. The sparks will probably be very short, about one-half inch long. If the machine is working properly the sparks should occur quite rapidly.

The amount of electricity which the machine will produce can be greatly increased, in fact sparks two or three inches long can be secured, by treating the surface of the leather rubber with an amalgam.

The amalgam is made by melting one ounce of tin and adding to it one ounce of zinc in small bits. When the zinc has melted remove the ladle from the fire and just before the mixture of metals is about to harden, add two ounces of mercury which has been previously warmed.

Do not inhale any of the vapor which will rise from the hot metal when the mercury is added. Pour the mixture into

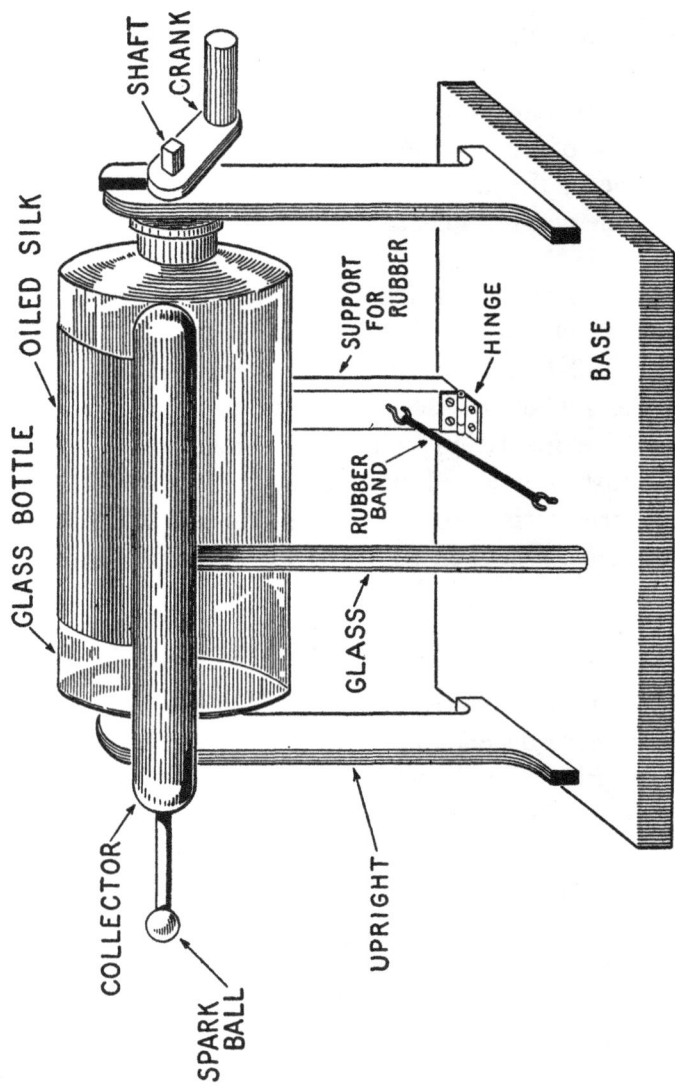

SHAFT

CRANK

OILED SILK

GLASS BOTTLE

SUPPORT FOR RUBBER

HINGE

BASE

RUBBER BAND

GLASS

COLLECTOR

SPARK BALL

UPRIGHT

FIG. 43. — The Complete Cylinder Electric Machine.

cold water. This will reduce the amalgam to small grains. Pour off the water and grind the grains of amalgam into a fine powder with a chemist's mortar and pestle.

The surface of the leather should be *thinly* smeared with lard and some of the amalgam powder rubbed on it.

The machine will not generate in summer or on damp winter days. But when the weather is cold and the atmosphere is dry, it will produce some powerful sparks at the prime conductor.

In order to obtain the best results from the machine, it must be dry and free from dust and particles of amalgam adhering to the surface of the bottle and the under side of the oiled silk. The insulating column which supports the prime conductor must be dry and free from dust. The machine will usually perform satisfactorily if it is warmed slightly near a radiator or stove and then rubbed with a warm woolen cloth.

A Wimshurst Machine

The static machine named after James Wimshurst, its inventor, is a far more dependable generator of electricity than the frictional machine just described. In principle, it is a form of revolving electrophorous and is probably the most efficient type of static machine.

Wimshurst machines were once used for operating X-ray tubes.

The Wimshurst machine consists of two varnished glass plates revolving in opposite directions. On the outside of each of these plates are cemented a number of tinfoil "sectors." Two movable conductors at right angles to each other extend obliquely across the plates, one at the back and the other at the front. These conductors, which are called neutralizers, each terminate in wisps or bunches of tinfoil which brush against the sectors as the plates revolve. The electricity on the whirling sectors is gathered by a set of collector pins arranged in a somewhat similar manner to those on the frictional machine.

Wimshurst machines have been built with plates varying from two inches in diameter to seven feet. The machine described below will give sparks three inches long. It can be made in a larger or smaller size by keeping the same relative proportions. At the time the Wimshurst machine was invented, the only material suitable for making the plates was glass. Since then, new materials such as Micarta, Bakelite laminated, etc., have been developed. From a mechanical standpoint, these materials are much superior to glass. They are very strong. They are expensive in comparison to glass and not easy to obtain. Glass plates are described and illustrated in this book but any strong insulating material which comes in sheet form may be used in place of glass. Even old phonograph records may be used.

The Plates are each eighteen inches in diameter. Purchase two panes of clear white window-glass twenty inches square from a glass dealer. The

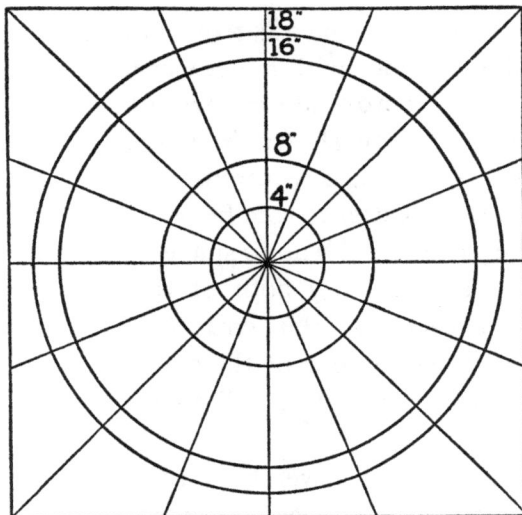

FIG. 44. — Paper Pattern for Laying Out the Plates and Locating the Sectors.

glass should be of the thickness known as "single light" and should be free from wavy spots, bubbles or other imperfections.

Many glaziers have a circular glass cutter and can cut two eighteen-inch circles for you and save you time and trouble. If you wish to cut your own "plates," this is the way to do it:

The work is first laid out in the form of a pattern on a piece of stiff paper twenty inches square. Using the exact center of the paper square as a center, describe a circle four inches in diameter. Using the same center, draw three other circles making them respectively eight, sixteen and eighteen inches in diameter. Then mark sixteen radial lines, from the center, making them equal distances apart. Lay one of the glass panes over the pattern and cut out a glass disk eighteen inches in diameter by carefully tracing the outline of the large circle on the pattern. Use a new glass cutter and dip it into kerosene from time to time.

When you have cut out two disks or plates, rub the sharp edges smooth with a wet Carborundum stone.

The Sectors are cut from heavy tinfoil. The shape and dimensions are shown in one of the illustrations. Thirty-two such sectors are required. The easiest way to make them is to cut out a pattern from heavy cardboard, hold the tinfoil against the pattern and trim to shape with a pair of scissors.

Clean and dry both of the glass plates carefully and give them each two *thin* coats of white shellac. After the shellac has dried, lay one of the plates on the paper pattern so that the outside of the plate will coincide with the largest circle on the paper.

Then place a weight in the center of the plate so that it will not slip out of place on the pattern, and stick sixteen of the tinfoil sectors onto the plate with thick shellac. The sectors are arranged symmetrically on the plate, using the eight-inch and the sixteen-inch circles and the sixteen radial lines as guides.

Both plates should be handled in the same manner. Each sector should be carefully pressed down on the glass, so that it is cemented firmly in place without air-bubbles or creases.

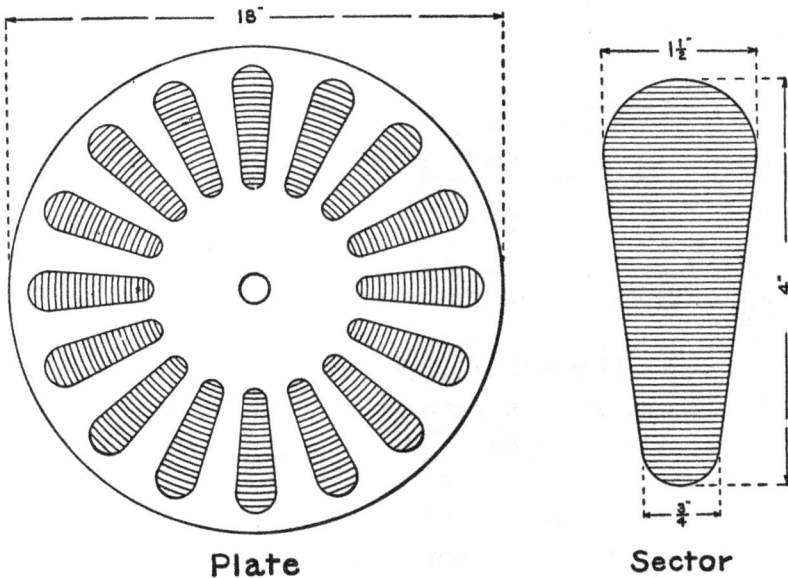

Plate   **Sector**

FIG. 45. — Plate with Sectors in Position, and a Pattern for the Sectors.

When all the sectors are fastened in place the plates should appear as shown in the illustration.

The Bosses are a combination of pulley and bearing used to support the plates. They must be turned out on a wood turning-lathe. The large end is four inches in diameter and the small end one and one-half inches in diameter. The groove turned in the small end of each boss serves as a pulley to accommodate a round leather belt.

A hole should be drilled into the axis of each boss about half-way through from the small end. These holes are bushed with a piece of brass tubing having an inside diameter of one-half inch. The tubing should go into the hole very snugly and be a "driven fit."

The bosses should both be given a coat of shellac, and after this is dry, fastened to the glass plates on the same side to which the tinfoil sectors are attached. If the plates are made of phonograph disks or a material such as Micarta, they may be fastened to the bosses with flat-head screws, but glass plates must be cemented. Use any one of the several varieties of cement made for mending broken china and glassware.

Brass bushing

groove

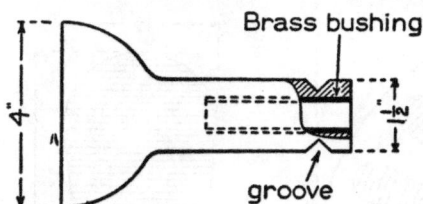

FIG. 46. — A Side View of One of the Bosses, Showing the Brass Bushing Used.

The Frame or Base is composed of two strips of wood fastened to cleats or battens. The dimensions and construction are illustrated.

The Uprights are notched so that when they are fitted into place on the frame the bottom of each will just clear the surface of a table. The machine is designed so that the cross pieces on the frame form its support. If the uprights are too long the machine will not rest firmly.

The Driving Wheels are turned out of wood on a lathe. They are seven inches in diameter. A groove is turned in the edge to carry a small leather belt. The wheels are glued to a wooden shaft made from a piece of dowel rod, one inch in diameter. The ends of the shaft pass through holes in the uprights, five inches above the bottom. The front end of the shaft is fitted with a crank and a handle.

The plates and bosses are supported on short iron shafts passing through the top of the uprights into the brass bushings. One end of each shaft is filed flat where it passes through the upright so that it may be firmly held by a set-screw and prevented from turning. Cement a small fiber washer or disk to the center of one of the glass plates so that it will prevent them from rubbing together when revolving.

The Collectors, Quadrant Rods, etc., are mounted on glass rods or tubes, preferably one inch in diameter. Water gauge

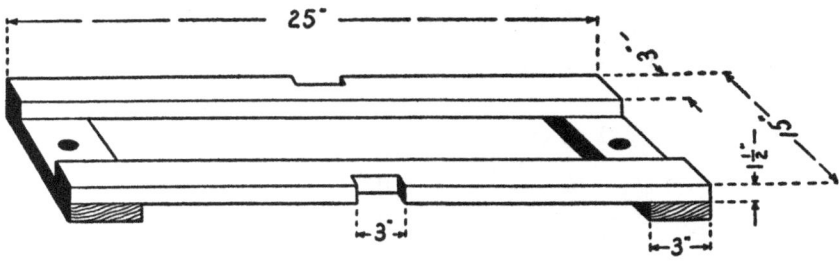

FIG. 47. — The Frame of the Wimshurst Machine.

tubing may be used but will not be as strong because of its smaller diameter. The bottoms of these insulating tubes or rods are cemented into holes in the wooden frame.

The illustrations show the upper ends of each rod to be fitted with a brass ball two inches in diameter. This is the most desirable method of constructing the machine because it avoids the leakage of static electricity at a place where it will try to escape if there are sharp angles, points or edges. Large brass balls are difficult for

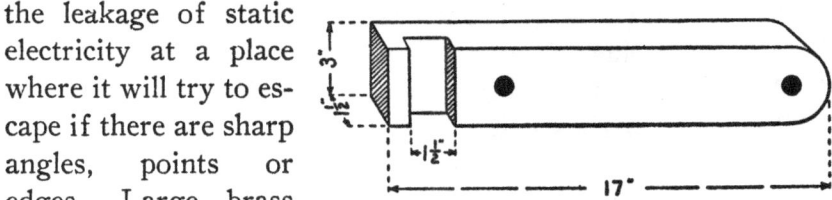

FIG. 48. — The Upright.

the average boy to make or obtain but there is a way to overcome the difficulty. Turn the balls out of wood, sandpaper them smooth, rub with graphite and then electroplate them with copper. Instructions for electroplating are given in a later chapter.

The rods should be of the proper length to bring the center of the balls on a line with the center of the plates.

Make two collector forks by bending brass rod, three six-teenths of an inch in diameter, and solder small brass balls on the ends. The collectors should be about eleven inches long.

The collector points are made by cutting off the sharp ends of dressmaker's pins and soldering them in place.

Side view   End view

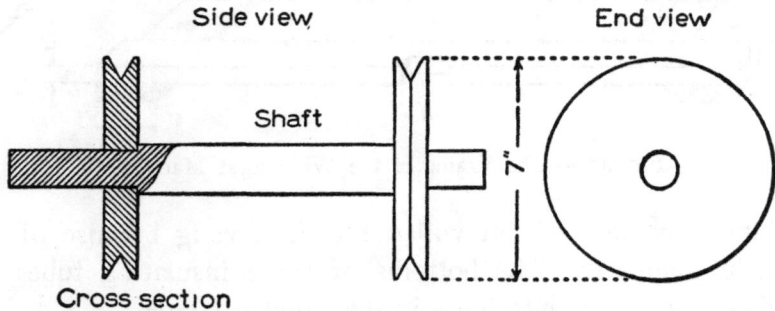

Shaft

Cross section

FIG. 49. — The Driving Wheels and Axle.

Bore a horizontal hole through each of the large brass balls supported by the glass rods and pass the shanks of the collectors through and solder them in place. The ends of the collector shanks are each fitted with a brass "discharge ball."

The quadrant rods extend over the tops of the plates, and are made of brass. They are loose so that they may be moved about or removed entirely.

A small brass ball three-quarters of an inch in diameter should be soldered to the top of one of the quadrant rods and a similar ball two inches in diameter to the other.

Two brass balls, two inches in diameter, are fitted over the ends of the shafts which project through the uprights. Bore a one-quarter inch hole through each ball at right angles to the axle and slip a one-quarter inch brass rod through and solder it fast. The ends of the rods should be tipped with a bunch of

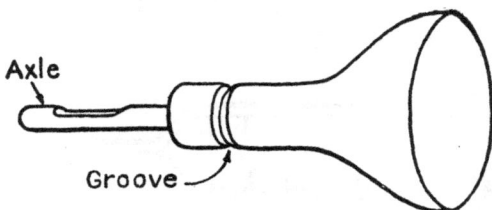

FIG. 50. — The Boss and Axle.

tinsel or fine copper wires and be curved so that the brushes will just touch the sectors on the plates.

These are the neutralizers and are arranged in the approximate positions shown in sketch of the finished machine.

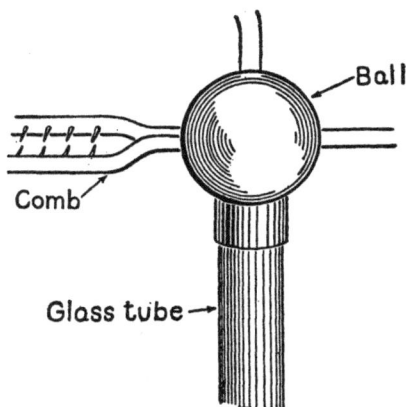

FIG. 51. — The Ball, Comb, etc., Mounted on the Glass Rod.

The driving wheels turn the plates through two round leather belts. The belt at the rear of the machine is crossed in order to make the plates revolve in opposite directions.

It may be necessary to charge the machine for the first time it is used by touching several of the sectors with the charged cover of an electrophorus. Then, if the handle is turned, the machine will start to generate and the electricity will discharge across the spark-gap at the top of the machine in the form of bright blue sparks.

EXPERIMENTS WITH A STATIC MACHINE

An almost unlimited number of interesting experiments can be performed with a static electric machine. There is only space enough to describe a few. Others may be found in old text books on physics.

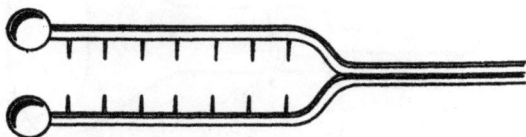

FIG. 52. — A Comb or Collector.

A Leyden Jar is a means of accumulating and storing static electricity for a short time. It consists of a glass jar partly coated with tinfoil inside and out. A small brass chain attached to the end of a brass rod connects with the inner coating. The rod passed through a wooden cover fitted to the jar and usually terminates in a brass ball.

It is not difficult to make a Leyden jar. Usually the only trouble involved is to obtain a jar having the proper quality of glass.

The "hard" glass used in chemical glassware is very satisfactory.

A small Leyden jar may be made from a test tube. Larger sizes may be made from beaker glasses.

The jar must first be thoroughly cleaned and dried. The inside is then given a coating of shellac. The tinfoil coatings should cover only the lower half or two-thirds of the jar. Before the shellac on the inside of the jar is dry, insert the tinfoil and press it smoothly against the glass. The outside of the jar is treated and coated in the same manner. The inside and outside of the bottom are also coated by cutting the tinfoil in circular pieces and shellacking them in place.

To charge a Leyden jar from the frictional machine, grasp

FIG. 53. — The Complete Wimshurst Electric Machine. B B B B, *Brushes.* C C, *Combs.* D B, *Discharge Ball.* I I, *Glass Rods.* H, *Handle.* Q. Q, *Quadrant Rods.* S S S S S, *Sectors.* S. G, *Spark-Gap.* P P, *Driving-Wheels.*

For the sake of clearness, several of the sectors are not shown.

it in the hand near the bottom and hold the knob against the prime conductor.

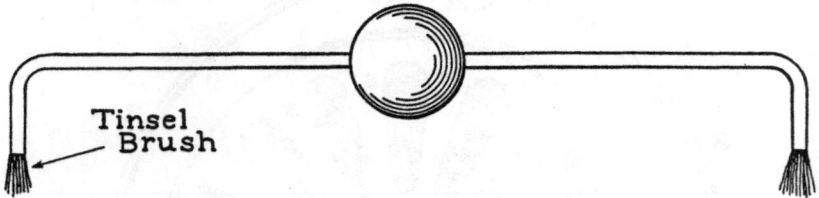

FIG. 54. — The Neutralizing Rods with Tinsel Brushes on Ends.

A small Leyden jar can deliver a stinging shock, a shock powerful enough to make you drop the jar, so be careful. Don't touch the brass knob or anything to which it is connected.

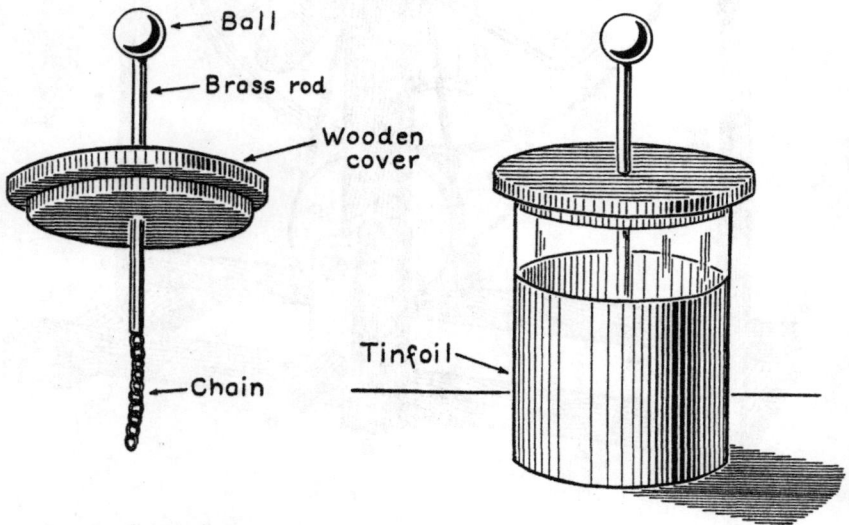

FIG. 55. — The Leyden Jar.

A Leyden jar may be charged from a Wimshurst machine by connecting the tinfoil coating on the outside of the jar to one of the discharge balls. The inside coating is connected to the other discharge ball.

You can discharge a Leyden jar by bringing a piece of wire which is connected to the outside coating, near the ball on the rod. When the wire is close to the ball, the electricity will jump across the space in the form of a snapping white spark.

You can make a discharger from a piece of brass wire bent into the form of a large "U" and fitted with a handle.

Igniting Powder. Bore a hole one-half inch in diameter and about one inch deep in a small block of hardwood. Force two small brass wires through holes in the side of the block until the ends are about one-eighth of an inch apart. Put a pinch of flashlight powder in the "mortar." Tie a piece of wet cotton twine, three inches long, to one of the wires and attach it to the outside coating of a charged Leyden jar. Connect the knob of the jar to the other wire. The powder will immediately flash. It should not be necessary to advise keeping the face and fingers away from the mortar to avoid the possibility of a burn.

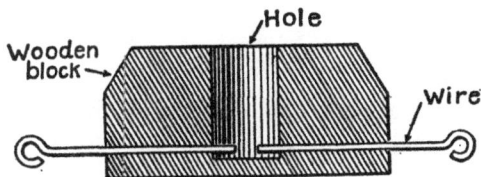

FIG. 56. — A Wooden Mortar for Igniting Gunpowder.

The purpose of the wet string is to force the Leyden jar to discharge slowly. The spark from a Leyden jar is ordinarily so sudden and short that it would blow the flashlight powder out of the mortar without igniting it.

Brush Discharge. If you operate a static machine, it does not matter whether it is the frictional or Wimshurst type, you will be able to see electricity leaking away or escaping in the form of a pale blue or violet colored light. This is called *corona* or *brush* discharge. It occurs at points, corners and edges.

The discharge of electricity from a sharp point is also called a convection discharge.

The convection discharge from a sharp pointed wire connected to the prime conductor of the frictional machine will
blow the flame of a candle
held near it.

Wire connected
to static machine

FIG. 57.— A Strong Static Discharge
from a Sharp Point Will Blow a
Candle Flame.

The Electric Whirligig
consists of an S-shaped piece
of brass or copper wire,
pointed at both ends and
supported on an upright pin
or needle. The depression
on the underside of the
whirligig which enables it to revolve on the pivot is a small
hole drilled part way into the wire with a tiny drill.

The force of a convection discharge reacts from the point
where it takes place and tries to push the point away. It is like
the "kick" or recoil of a gun. It is the reactive force of the discharge which drives the whirligig.

The needle is stuck head down in a cork in the top of a bottle
and connected with the prime conductor of the static machine.

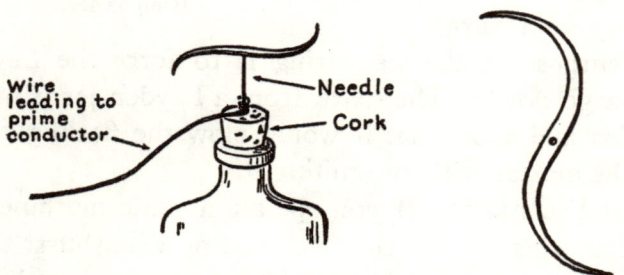

Wire
leading to
prime
conductor

Needle

Cork

FIG. 58.— The Electric Whirligig.

When the latter is set in motion, electricity will escape from the
points and the whirligig will revolve at a high rate of speed.

Electric Umbrella. The repulsion of similarly electrified bodies which was illustrated by the action of the pith ball electroscope may be better illustrated by pasting some narrow streamers of tissue paper one-sixteenth to one-eighth of an inch wide to a small tinfoil covered cork. The cork is mounted on the upper end of a stiff copper wire supported by a bottle. When the wire is connected to a static machine, the tissue paper strips will become charged and spread out like an umbrella.

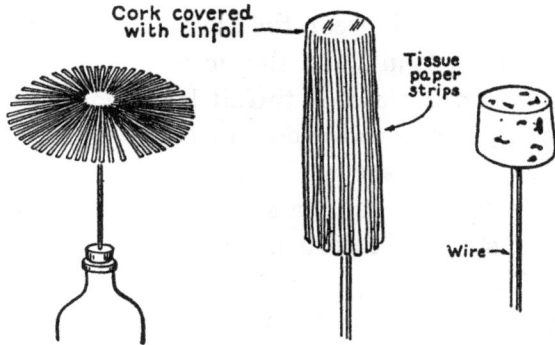

Cork covered with tinfoil

Tissue paper strips

Wire

FIG. 59. — An Electric Umbrella.

A Lightning Board. A very surprising effect can be produced by passing the spark from a Leyden jar or a static machine through a "lightning board."

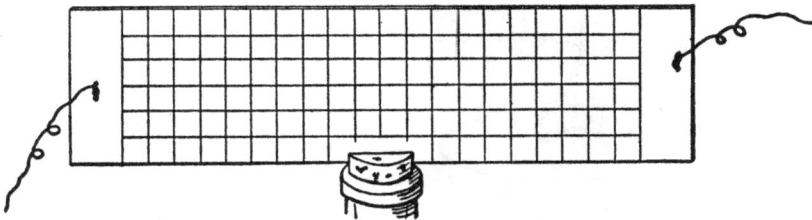

FIG. 60. — A Lightning Board.

A "lightning board" consists merely of a pane of glass having a number of small squares of tinfoil stuck on it so that when the electrical discharge passes over it, sparks take place between the little tinfoil squares and produce an effect somewhat like miniature lightning.

A "lightning board" may be made from a strip of ordinary window glass about nine inches long and two inches wide.

Thoroughly clean the glass strip and give it a coat of shellac varnish. While the shellac is in the "tacky" or sticky stage, press on a piece of tinfoil large enough to cover one side of the glass and rub it down smoothly.

After the shellac is dry cut the tinfoil up into innumerable little squares, using a safety razor blade as a knife. Leave two solid strips of tinfoil at the ends of the pane.

The lightning board is mounted on an insulating support by cementing it in a slot in the cork of a bottle. Connect one of the tinfoil strips to the prime conductor of the frictional machine and the other to the earth or body. If you use a Wimshurst machine connect the strips to the discharge balls.

An Electrical Dance. This is an experiment which illustrates attraction and repulsion.

A number of small balls of cork or pith are enclosed in a

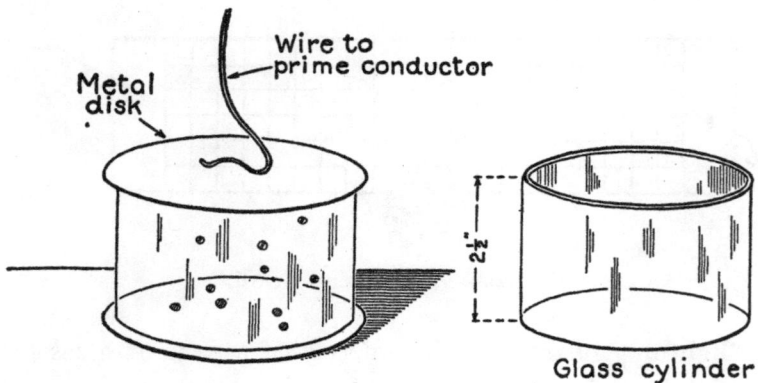

FIG. 61. — An Electric Dance.

cylinder of glass about two and one-half or three inches high, formed by cutting off the bottom of a small jar. The top and bottom of the cylinder are closed by two circular pieces of sheet metal. The top disk is connected to the prime conductive of a fractional machine or the positive discharge ball of a Wimshurst machine. The bottom disk is connected to the rubber of the frictional machine or the negative discharge ball of a Wimshurst machine.

When the machine is set in operation, the little balls will dance up and down. Bits of feather or small pieces of paper, cut to represent figures of men and women, may be used as well as pith or cork balls.

Positive and Negative Discharge. The appearance of the brush produced by a convective discharge of electricity depends upon whether the charge is positive or negative.

The brush from a sharp point connected to the prime conductor of a fractional machine is fan shaped because the charge is positive.

The rubber is negative but you cannot secure negative electricity from it because it is not insulated, and leaks away.

You can however secure a negative discharge from one of the discharge balls of a Wimshurst machine and

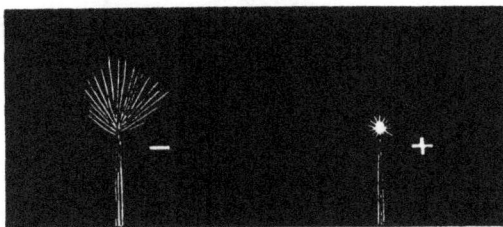

Fig. 62. — A Negative Discharge from a Point Forms a Brush or Spray. The Positive Discharge is a Bright Star.

if you examine the brush in a dark room, it will be star-shaped. Michael Faraday explained this a hundred years ago. He showed that a negative charge escapes more readily into the air than an equal positive charge.

Lichtenberg's Figures illustrate the fact that one coating of a Leyden jar is charged with positive electricity and the other with negative.

Hold a jar charged from the prime conductor of a frictional machine in the hand and draw a circle on a cake of resin or

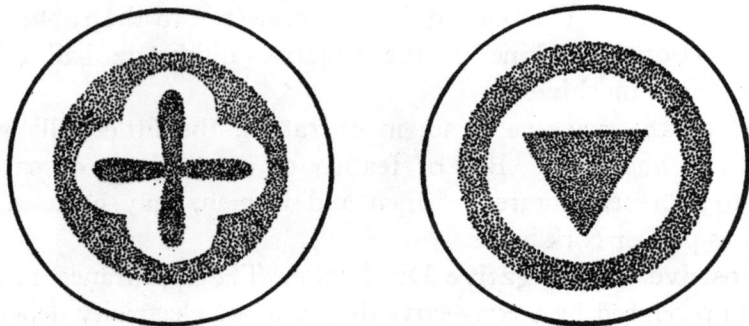

FIG. 63. — Lichtenberg's Figures.

sheet of hard rubber with the knob of the jar. Then, the jar having been placed on an insulator and picked up by the knob, draw a cross on the resin or hard rubber with the outer tinfoil coating.

If, now, a mixture of red-lead and powdered sulphur is dusted through a piece of muslin and allowed to fall upon the resin, Lichtenberg's figures will form. The sulphur will attach itself to the circle drawn with positive electricity and the red-lead to the negative cross.

When red-lead and sulphur are mixed, the sulphur becomes negatively electrified and the red-lead positively. The sulphur will arrange itself in tufts with numerous diverging branches, while the red-lead will take the form of small circular spots.

Chapter IV

VOLTAIC CELLS AND BATTERIES

THE static electricity generated by an electrophorous, static machine or rubbing glass, sealing wax, etc., is of no *practical* use. A continuous current of electricity is needed to drive motors, furnish light, carry messages and do other useful things.

There are several means of generating a continuous current, the two principal ones being:

A wire moving in a magnetic field as in the dynamo, and

Chemical action as in a Voltaic cell.

For current to drive their various electrical devices, "boy electricians" depend upon batteries, a small dynamo, or the house-lighting current.

All homes and workshops are not supplied with electric current. Small dynamos require a source of power such as an engine or water wheel. Another chapter explains how the 110 volt current may be used safely for experimenting. For many purposes it is not suitable without transforming, rectifying, and filtering.

For experimenting with spark coils, telegraph instruments, telephones, electromagnets, etc., use batteries.

You can buy first class dry cells of the size known as No. 6 for twenty-five cents each, and these will prove to be the most satisfactory source of battery-generated electric current.

If you figure out what it costs to buy zinc, chemicals, etc., you will learn that it is not economical to build your own cells. The real value in making your own batteries will not lie in the fact that they are cheaper than factory-made cells, but in what you *learn* in making them.

Build your own batteries first. Then, after you have learned how they are made and something about their proper care, use factory-made dry cells or storage batteries.

Many hundreds of thousands of experiments have been made by skilled chemists and physicists in an effort to devise better Voltaic cells. There are many different kinds of cells but none of them is perfect. A number of different types are described in this chapter.

The words Voltaic cell are used as a name for the means of producing a current of electricity with chemicals.

The Voltaic Cell is called after its inventor, Alessandro Volta, a professor in the University of Pavia, Italy.

Cells are usually considered *one* element of a *battery*. A

FIG. 64. — The Voltaic Cell.

cell means only one, while a battery is a group of cells. It is not a proper use of the word to say "battery" when only *one* cell is implied. This is a very common error.

A simple Voltaic cell is made by placing some water mixed with a little sulphuric acid in a glass tumbler and immersing therein two clean strips, one of zinc and the other of copper. The strips must be kept separate from each other. The sulphuric acid is diluted by mixing with about ten times its volume of water. When

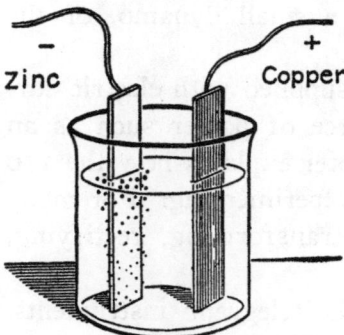

mixing acid and water, always remember to pour acid into water. Never pour water into acid.

The copper and zinc strips are called the *elements*. The water and acid are the *electrolyte*.

A copper wire is fastened with a screw and nut, or by soldering to the top of each of the elements, and care must be exercised to keep the wires apart. As has been said, the zinc and copper must never be allowed to touch each other in the solution, but must be kept at opposite sides of the jar.

The sulphuric acid solution attacks the zinc, causing it to disappear. The zinc changes into zinc sulphate and goes into solution in the electrolyte. This chemical activity produces an electric current, which in the case of a single small cell will be found to be sufficient to ring a bell or buzzer or run a very small toy motor.

Bubbles rise from the zinc. This is the gas called hydrogen and indicates that a chemical action is taking place. As the zinc changes into zinc sulphate, hydrogen is set free from the sulphuric acid. It will be noticed that apparently no bubbles arise from the copper plate.

The Voltaic cell may be likened to a furnace in which the zinc is the fuel which is burned to furnish energy. We know that when fuel is burned it gives out energy in the form of heat. When one metal is "burned" or consumed in a chemical solution in the presence of another metal it gives out energy partly in the form of heat and partly in the form of an electric current. The acid electrolyte may be likened to a fire and the copper element to a ladle which dips into the cell to pick out the electricity but takes no part chemically.

If the wires connected to the zinc and copper elements are touched to the tip of the tongue, a peculiar salty taste will be produced. This is the effect of a current of electricity upon the nerves of taste.

If a cell is made of two zinc plates or two copper plates, no electricity will be produced. In order to produce a current, the electrodes must be made of two different metals upon which

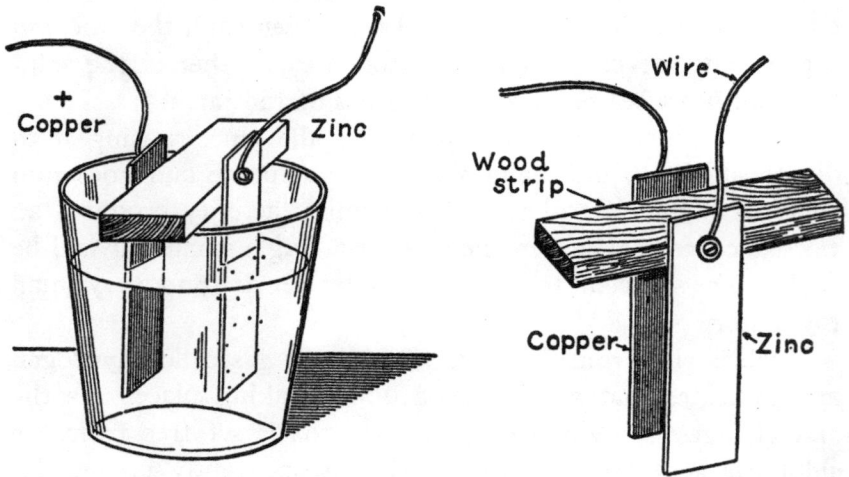

FIGS. 65 and 66. — The Elements of a Simple Home-Made Voltaic Cell.

the electrolyte acts differently. Therefore, in order to make a Voltaic cell it is necessary to have present a metal which may be consumed, a chemical to consume or oxidize it, and an inactive element which is merely present to collect the electricity.

When the two wires connected to the two elements are joined together, a current of electricity flows from the copper through the wire to the zinc. Inside the cell, in the electrolyte, the current flows from the zinc to the copper. The copper is known as the *positive* pole and the zinc as the *negative*.

Name of Cell	Positive Element	Negative Element	Exciting Fluid	Depolarizing Fluid	E. M. F. in Volts	Remarks
Bunsen	Carbon	Zinc	Sulphuric acid, dilute	Nitric acid	1.734	The carbon and nitric acid are placed in a porous cup.
Bunsen	"	"	"	Chromic acid	1.734	The carbon and chromic acid are placed in a porous cup.
Chromic acid, single fluid	"	"	Sulphuric acid and chromic acid, dilute, mixed	None separate	2.2	
Daniell	Copper	"	Zinc sulphate solution	Copper sulphate solution	1.079	The zinc is placed above the copper and the solutions are separated by gravity forming the so-called "gravity cell."
Fuller	Carbon	"	Chloride of zinc solution	Bichromate of potash and hydrochloric acid	1.5	The carbon and depolarizer are placed on the *outside* of a porous cup.
Niaudet	"	"	Common salt solution	Chloride of lime	1.5	The carbon and chloride of lime are placed in a porous cup.
Poggendorf	"	"	Saturated solution of bichromate of potash and sulphuric acid	None separate	1.98	

How to Make a Voltaic Cell

A Voltaic cell may be made by cutting a strip of sheet zinc and a strip of sheet copper, each three and one-half inches long, and one inch wide. A small hole bored through the upper end of the strips will permit them to be mounted on a wooden strip with a small screw as shown in the illustration. The connecting wires are placed under the heads of the screws or soldered to the tops of the strips. The screws used for mounting the strips should not be exactly opposite, or the points may touch and short-circuit the cell.

An ordinary tumbler or a jelly glass will make a suitable jar for the electrolyte. The electrolyte should be composed of

One part of sulphuric acid

Ten parts of water.

The cell will ring an electric door bell vigorously or run a small motor. Two cells, connected in series to form a battery, will light a small flashlight lamp.

The original form of Voltaic cell, consisting of zinc and copper elements and sulphuric acid electrolyte, has two disadvantages. Ordinary zinc contains impurities, and largely on that account the zinc is consumed by the acid no matter whether the electric current is being used or not. This is called *local action* and it is very wasteful of the chemicals. It can be partially prevented by a method which will be described later but it cannot be entirely eliminated.

The other disadvantage is that the cell becomes "tired." When the cell is used, its strength falls off rapidly. This is due to an action called *polarization* and is caused by small bubbles of hydrogen gas which collect on the copper element.

There are a number of different chemicals which may be added to the electrolyte to avoid polarization and there are also a number of different electrolytes and different elements that make a more satisfactory cell than Volta's original creation.

THE LECLANCHE TYPE CELL

Before dry cells became as efficient as they are today, a type of cell called the Leclanche, after its inventor, was commonly used for telephones, bell-ringing and "open-circuit work." Open circuit work means where current is required occasionally, as distinguished from closed-circuit work where current is required constantly.

The elements of a Leclanche cell consist of zinc and carbon. The carbon is the positive element and replaces the copper of the original Voltaic cell. The electrolyte is a solution of *sal ammoniac* or ammonium chloride solution and does not actively attack the zinc when no current is drawn from the cell. The Leclanche cell has an E. M. F. (for explanation see Chapter VI) of 1.4 volts which is about one and one-half times that of Volta's cell.

If current is drawn from a Leclanche cell for any appreciable length of time hydrogen gas collects on the carbon element and the cell becomes polarized. When polarized, it will not deliver much current. However, it will recover if given an opportunity to rest.

The original Leclanche cells were provided with a chemical depolarizing agent. The carbon element was in the form of a flat plate placed in a porous earthenware cup filled with manganese dioxide.

Chemists class manganese dioxide as an oxidizing agent, which means that it will furnish oxygen with comparative ease. Oxygen and hydrogen have a strong chemical affinity or attraction for each other. When the carbon element is packed in manganese dioxide any hydrogen which collects on the carbon is immediately seized by the oxygen of the manganese dioxide and chemically united with it to form water.

HOME-MADE BATTERIES

The carbon elements for making experimental cells are most easily obtained from old dry cells. A dry cell can be broken open with a cold-chisel and a hammer. Be careful not to break the carbon.

Ordinary jelly-glasses make good jars for small cells. Fruit-jars may be used for large batteries. Cutting the tops off will increase the size of the opening.

When the carbon rod from an old dry cell is too long it can be cut to the proper length with a hacksaw. The lower end is the part which should be discarded because the top carries a binding-post.

The zinc element for home-made cells is somewhat of a problem. At one time, all electrical stores carried zinc battery rods in stock. Now they are hard to find. So is sheet zinc, once much used by plumbers and sheet metal workers. Other materials such as copper and heavily galvanized iron have largely taken its place.

Sheet zinc may still be found in some of the larger hardware stores and plumbing shops. From it, strips suitable for home-made batteries may be cut. Zinc melts at 787 degrees Fahrenheit. Since this is a temperature easily reached in the flame of a gasoline blow torch or an ordinary cooking stove, scrap zinc may be melted and cast in the form of rods or plates. Plaster of paris molds may be used. They must be thoroughly dry.

Chemically pure zinc is not attacked by dilute sulphuric acid but if you place a piece of "commercial" zinc in dilute sulphuric acid, bubbles of gas will be given off, more particularly at certain parts of the surface. This is "local action," and can be likened to a small cell, the impurity acting as the copper plate and the zinc around it being eaten away. Ordinary commercial zinc is rapidly dissolved by dilute acid. The current thus generated is wasted as it circulates locally in the liquid. Local action can be prevented by *amalgamating* the zinc with mercury. This is readily done by dipping the zinc in dilute acid, and while wet rubbing a drop of mercury over its surface. The mercury will spread evenly over the surface and by its chemical action eliminate the effect of the impurities. Zinc, when amalgamated, acquires the properties of perfectly pure zinc.

It is necessary to support the carbon and zinc elements of a cell in such a manner that they cannot touch each other. If the support also covers the opening in the jar, it will prevent rapid evaporation of the electrolyte. Wood may be used for the purpose. It is a good plan to paint the wood with hot paraffin wax or, better still, soak it in hot paraffin until it is thoroughly impregnated.

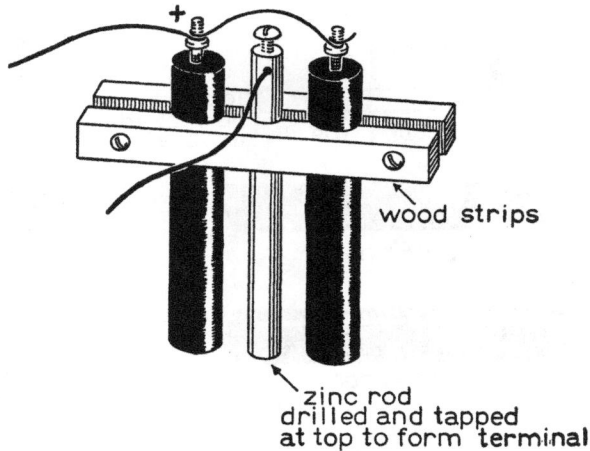

FIG. 67. — A Method of Making a Cell Element from Carbon Rods.

This prevents the electrolyte from "creeping" and soaking into the wood where it can form a current leak.

A simple Leclanche type cell may be made from a single carbon rod and a strip of zinc. If, however, the zinc element is placed in the center and from two to four carbon rods arranged around it, the cell will be more powerful. All the carbon rods should be connected together. A strong solution (about four ounces dissolved in a quart of water) of ammonium chloride, or, as it is more commonly called, sal ammoniac should be used as the electrolyte.

THE POGGENDORF CELL

The first toy electric railways manufactured in this country were operated by batteries consisting of zinc and carbon ele-

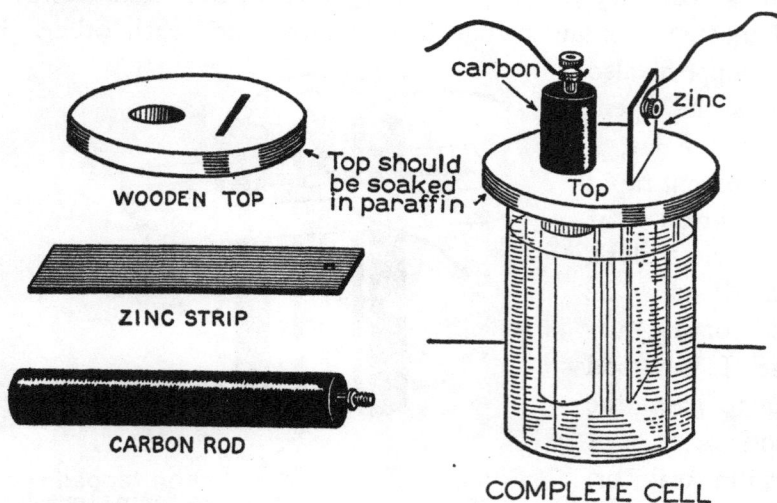

WOODEN TOP

Top should be soaked in paraffin

carbon

zinc

Top

ZINC STRIP

CARBON ROD

COMPLETE CELL

FIG. 68. — The Complete Cell.

ments and an electrolyte made by mixing "Electric Sand" or "Electropoian Fluid" with water. These mysterious names were the disguise for a sulphuric acid and dichromate of potash compound.

Dichromate of potash or dichromate of soda is a depolarizer. When either of these substances is added to a sulphuric acid solution, using zinc and carbon elements, it forms a very powerful cell, having an E. M. F. of about two volts.

A battery solution of this kind may be prepared by adding four ounces of dichromate of potash to a solution composed of four ounces of sulphuric acid mixed with sixteen ounces (one pint) of water.

It might be well at this time to caution the experimenter against the careless handling of sulphuric acid. It is not dangerous if handled properly, but if spilled or spattered carelessly it is capable of doing considerable damage to most things with which it comes in contact. Do not use sulphuric acid or sulphuric acid batteries except in a cellar workshop. Coming into contact with any organic matter such as woodwork, clothing, carpets, etc., it will not only discolor but eat such substances.

The best way to counteract the effects of acid which has been spilled or spattered is to flood it with water. An ammonia solution will neutralize acid and sometimes restore the color to clothing damaged by acid.

PLUNGE BATTERIES

All sulphuric acid cells have the objection that it is impossible to leave the elements in the electrolyte without wasting the zinc when the cell is not in use.

Batteries arranged so that the elements can be conveniently removed from the solution are called "plunge" batteries. The usual way to arrange a plunge battery is to fasten the elements to a chain or cord passing over a windlass fitted with a crank so that when the crank is turned the elements may be raised or lowered.

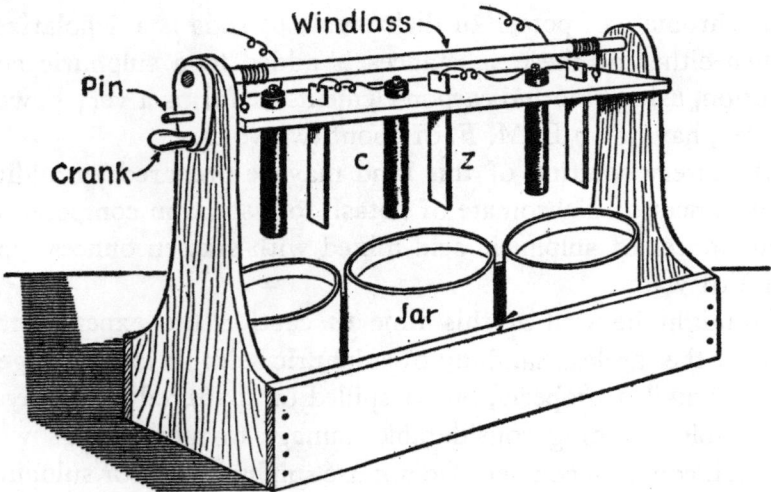

FIG. 69. — A Plunge Battery with Windlass.

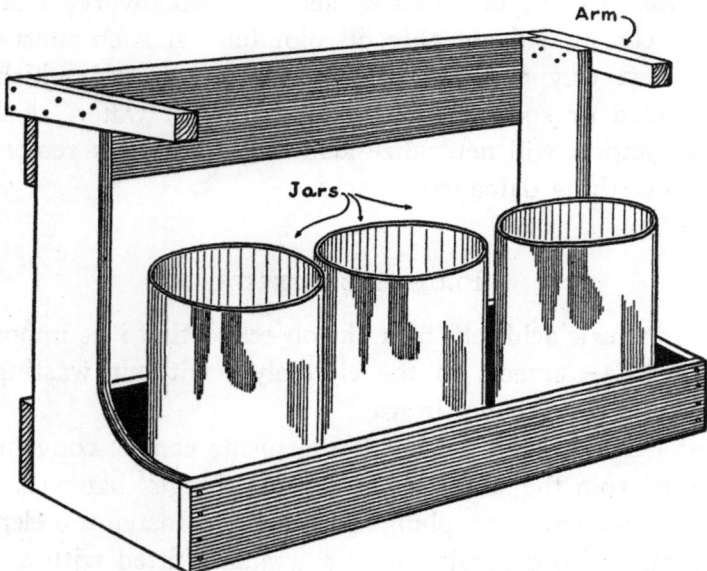

FIG. 70. — A Plunge Battery for Use with a Set of Elements as shown in Fig. 68.

A plunge battery of this sort is illustrated. The construction is so plainly shown that it is hardly necessary to discuss the details. The crank is provided with a dowel-pin which passes through a hole in the frame, so that when the elements are lifted out of the solution the pin may be inserted in the hole and the windlass prevented from unwinding.

A simpler method of accomplishing the same purpose is to arrange the cells in a wooden frame as shown in Fig. 70. The elements may be raised up out of the jars and laid across the two "arms" to drain.

A Tomato-Can Battery

This is a simple form of home-made cell whose principal fault is the low voltage that it delivers. It is liable to polarize but

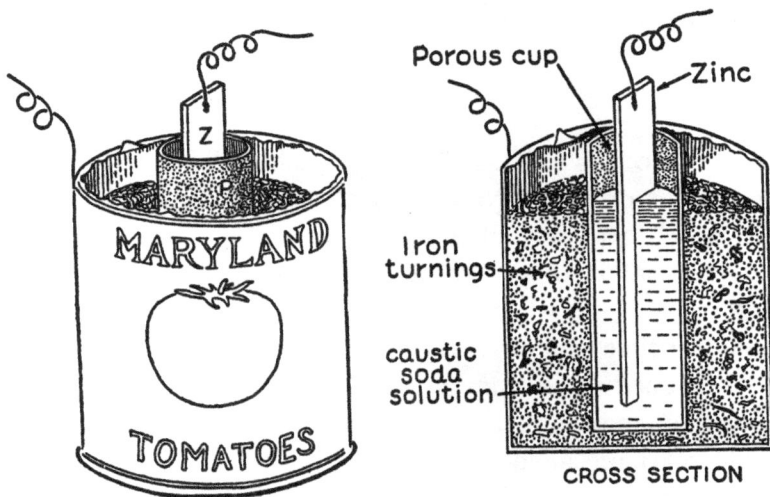

FIG. 71. — A Tomato-Can Battery.

the large surface of its positive element protects it to some extent.

The positive element and the receptacle which contains the

electrolyte is a tomato can. Within it is a small unglazed flower pot which serves as a porous cup. The hole in the bottom of the flower pot should be plugged with clay or wax.

The space between the can and the porous cup is filled with fine scrap-iron such as borings and turnings from a machine shop.

The negative element is a strip of zinc placed in the porous cup.

The electrolyte is a ten-per-cent solution of caustic soda. The level of the electrolyte should not be above the top of the flower pot.

The table on page 77 gives the names, elements, exciting fluid, depolarizer and voltage of the better-known types of cells. With this as a guide, the "boy electrician" can make cells other than those which have been described here in detail.

About Dry Cells

The dry cell is not, as its name implies, "dry," but the exciting agent or electrolyte, instead of being a liquid, is a wet paste which cannot spill or run over. The top of the cell is sealed so that it is always portable and may be used in any position.

The principle of a dry cell is the same as that of the Leclanche cell. The excitant is ammonium chloride, the elements are zinc and carbon, and the carbon is surrounded by manganese dioxide as a depolarizing agent.

Recharging Dry Cells is a subject that interests most experimenters.

Suggestions for "recharging" dry cells are sometimes published in the popular technical magazines but these schemes are more fantastic than practical.

The modern dry cell, when produced by a nationally known manufacturer, is the result of careful research and design. When the cell becomes exhausted, it is not because the moisture in the cell has dried up, or because any particular chemical contained in it has been consumed. All of the active ingredients of the cell are worn out. It has sort of died of old age.

Attempts to recharge the cell by passing a direct current through it as if it were a dry cell are of no avail. Neither will adding water or sal ammoniac solution help much.

An old cell may be given a short temporary new life by punching several holes in the zinc shell and then setting it in a glass jar containing a sal ammoniac solution. The cell should be allowed to remain in the solution until it will no longer deliver any current.

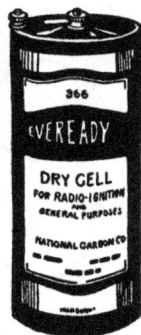

FIG. 72. — A Dry Cell.

CONNECTING CELLS

The wires used for connecting cells should not be smaller than No. 18 B. & S. gauge. Fine wire offers considerable resistance to the current and the full benefit of the batteries cannot be secured when it is used. The connections should be kept bright and clean. Tighten the binding posts with a pair of pliers.

If a number of similar cells be connected in *series,* the positive plate of one to the negative plate of the next, and so on, the difference of potential between the free ends will be increased in proportion to the number of cells so connected, but the combination cannot yield more current, or quantity of electricity, than a single cell would yield.

The other chief way of grouping is to connect all the positive poles together, and also all the negative; they are then said to be in *parallel*. Joined in this way we find that the voltage is

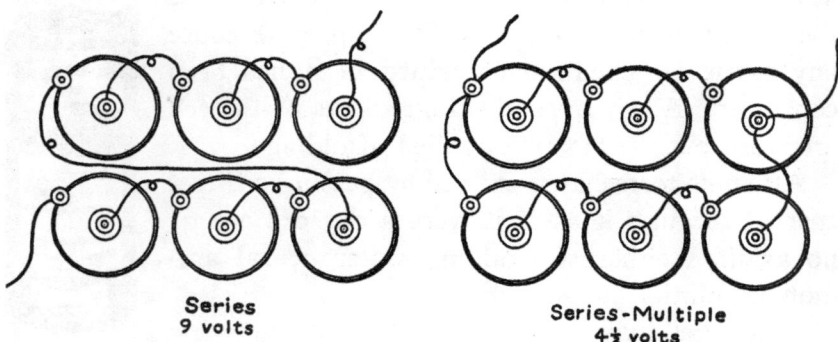

Series
9 volts

Series-Multiple
4½ volts

FIG. 73. — Two Methods of Connecting Cells so as to Obtain Different Voltage and Amperage Values.

the same as that of one cell, but arranged thus they are capable of yielding more current or a greater quantity of electricity.

MAKING A DRY CELL

There is no economy in making your own dry cells. The materials, when purchased in small quantities, cost as much as a completed factory-made cell.

However, for those who like to experiment, directions for building a simple but efficient dry cell will be found below.

Cut from sheet zinc as many rectangles, 8 x 6 inches, as it is desired to make cells. Also cut from sheet zinc an equal number of circles 2⅜ inches in diameter.

Roll the sheets up into cylinders, 2⅜ inches in diameter inside and six inches long. Overlap the edges and solder. Fit one of the circular pieces in one end of each of the cylinders and solder them securely in place, taking care to close all the seams and make them water-tight.

Procure some old carbon rods by breaking open old dry cells.

Make a wooden plunger about two inches longer than the carbon rod and three-quarters of an inch larger in diameter. Smooth it with fine sandpaper and give it two coats of shellac to prevent it from absorbing moisture. This wooden plunger is temporarily inserted in the center of one of the zinc cups and supported so that it will be about one-half inch above the bottom.

The electrolyte is prepared by mixing together the following ingredients in the proportions shown:

Sal Ammoniac	1 part
Zinc Chloride	1 part
Plaster of Paris	1 part
Flour	1 part
Water	2 parts

The mixture forms a paste which is firmly packed into the zinc shell around the wooden plunger, to a level about three-quarters of an inch from the top. The paste can be poured in very readily when first mixed but stiffens after standing a short while.

After the mixture has set, withdraw the wooden plunger, thus leaving a space inside of the dry cell somewhat larger than the carbon rod. The carbon is next inserted in this hole and the surrounding space is filled with a mixture composed of:

Sal Ammoniac	1 part
Zinc Chloride	1 part
Flour	1 part
Manganese Dioxide	3 parts
Powdered Carbon	1 part
Water	3 parts

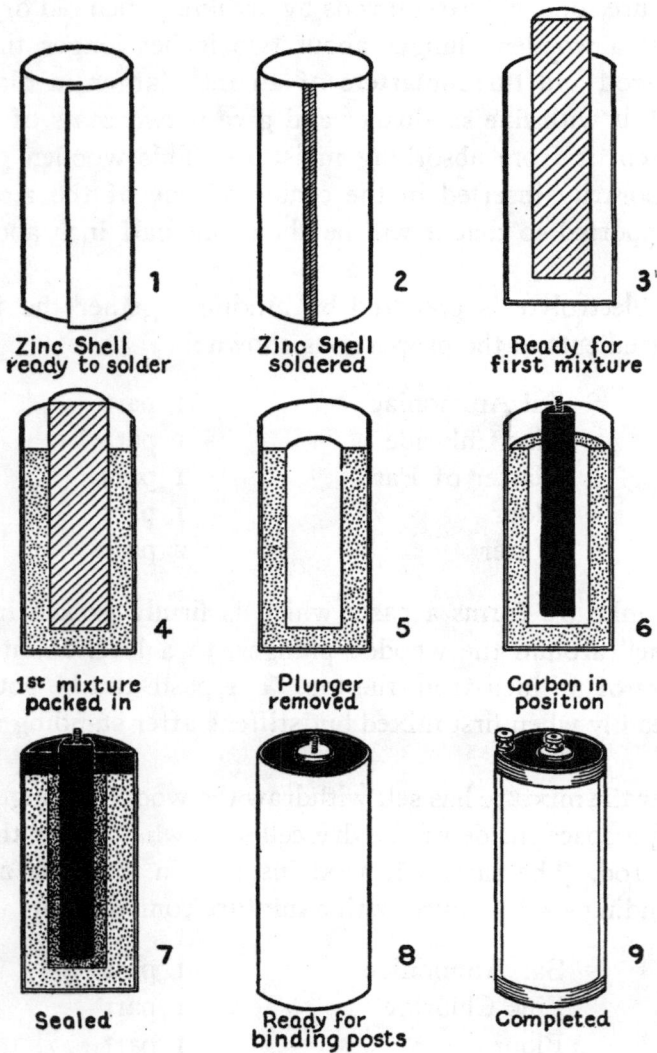

FIG. 74. — The Different Operations in Making
a Dry Cell.

The parts given in both of the above formulas are proportioned so that they can be measured by bulk and not by weight.

The granular carbon may be obtained by crushing an old dry cell carbon. The space remaining at the top of the cell should be poured full of molten asphalt, tar or pitch. Solder an 8–32 brass machine screw to the top edge of the zinc so that when fitted with a thumbnut it may be used as a binding post. Then when several thicknesses of heavy paper have been wrapped around the outside of the zinc case, the cell is ready to use.

So far in this chapter we have in every instance been discussing a type of cell which falls under the general classification of *primary* cell.

Storage or Secondary Cells

Storage or *secondary* cells differ from primary cells in that they will not give forth an electric current until they have been charged by passing an electric current through them.

What is really effected in a storage cell is the storage of *energy,* not the storage of electricity. The energy of the charging current changes into chemical energy and this chemical energy produces electricity when the cell is again discharged. There is really no more electricity in the cell when it is charged than after it is discharged. A storage cell is therefore a very convenient means of storing energy for future use.

Storage cells are made up of lead plates cast into a "grid" or framework. The spaces in the grid are filled with a paste of lead oxide. The plates are then "formed" by placing in an acid solution and connecting to a source of direct current.

The plates connected to the positive wire gradually turn dark brown in color, due to chemical changes in the paste which gradually turns into *lead peroxide*. The plates connected to the negative wire turn gray in color and the paste therein becomes a form of metallic lead called *spongy* lead.

After the plates have been formed they are assembled in a bundle. They are kept apart by means of strips of wood, perforated corrugated hard rubber or glass wool called separators.

The electrolyte used in a storage cell is dilute sulphuric acid.

When a cell has become exhausted, it is charged by connecting a source of direct current to its terminals and sending a current through it.

An Experimental Storage Cell

Two lead plates immersed in an electrolyte composed of ten parts of water and one part of sulphuric acid will demonstrate the action of a storage cell.

Mark one of the plates with an X to indicate that it is to be the positive. Connect this plate to the positive terminal of a battery having an E. M. F. of at least three volts. Connect the negative pole of the battery to the other lead plate.

Bubbles of gas will immediately begin to rise from the lead plates. Let the current flow for five or ten minutes and then disconnect the battery.

You will now find that your storage cell, for the two lead plates and the acid have become a storage cell, will ring a bell or run a small motor *for a few seconds*. The two lead plates became *charged* when the current from the battery passed through them.

A storage cell made of lead plates in the manner just described does not possess sufficient capacity to make it worth while as a practical cell. If, instead of solid flat plates, frameworks or grids filled with a paste of lead oxides are used, there is a considerable gain in the capacity of the cell.

A Six-Volt Battery

A storage cell, when charged, has an E. M. F. of two volts. Three charged cells, connected in series, have an E. M. F. of six volts.

A six-volt storage battery will provide a "boy electrician" with a most convenient and satisfactory source of electrical energy for experimenting. It is capable of delivering more current for a longer period than dry cells or other primary cells and is not expensive. A storage battery merely requires recharging and does not have to be thrown away each time it becomes exhausted.

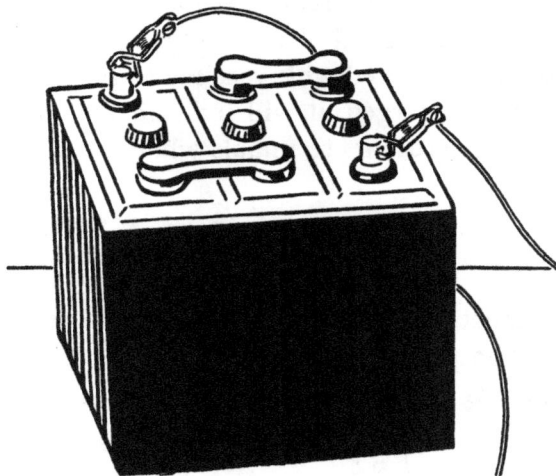

FIG. 75. — A Six-Volt Storage Battery.

Six-volt storage batteries manufactured for automobile starting and lighting can be bought for about four dollars. A second-hand battery, partially worn out and unsuitable for automobile starting, but perfectly satisfactory for experimenting, may be purchased at a junk shop for as little as 50 cents.

Storage batteries can be recharged with *direct* current only. A dynamo for recharging storage cells must be shunt wound. The voltage of the charging current must be greater than that of the battery. About three volts of charging current will be required for each cell of storage battery.

Alternating current may be used to recharge a storage battery when it has been first changed into *direct* current by a rectifier.

Storage cells should never be allowed to remain discharged or the plates will become hardened and "sulphated." A cell is never quite as good after it has been sulphated and it takes a great deal of charging and recharging to get the plates back into even fair shape.

Never short circuit a storage cell or discharge it too rapidly. It is a very good plan to keep the terminals of a storage cell or battery smeared with vaseline to reduce corrosion. When a cell is fully charged, it will indicate 2½ volts on a voltmeter connected across its terminals while the charging current is on.

How to Make a Storage Cell

The storage cell described below is made in a very simple manner, but has considerable energy-storing capacity.

The plates are cut out of sheet lead, preferably about one-quarter of an inch thick. They should be made to fit any small,

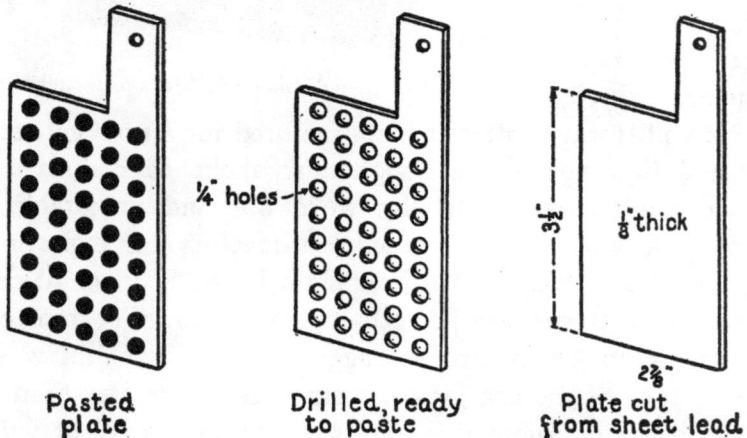

Pasted plate

Drilled, ready to paste

¼" holes

Plate cut from sheet lead

3¾"

⅛"thick

2⅝"

Fig. 76. — How to Make the Plates for a Storage Cell.

rectangular glass jars which the experimenter can obtain. A long terminal or lug is left projecting from each plate as shown in the illustration.

Small cells using sheet lead one-eighth of an inch thick and medicine bottles with the tops cut off as jars, built according to this plan make an excellent source of power for model boats.

Any number of plates may be placed in a single cell, depend-

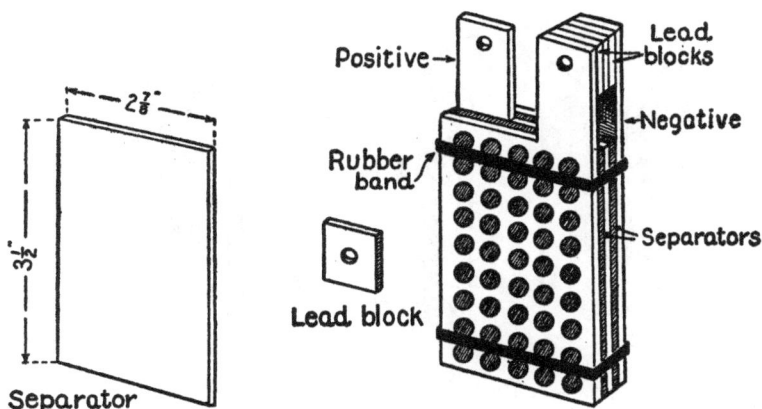

Fig. 77. — The Complete Elements for a Storage Cell.

ing upon the size of the glass jar. We will suppose that three will just fit the jar nicely.

Each cell will give two volts regardless of the number of plates. Increasing the number of plates, however, will increase the capacity and also the maximum discharge rate of the cell. Three cells (six volts) will form a convenient power supply for running small fan motors, spark coils, miniature lights, etc.

Cut out nine plates and pile them up in sets of three with a piece of thin wood (cigar-box wood) on the outside of each set. Clamp them in a vice and bore as many three-sixteenths inch holes as possible through each plate. Smooth off the burrs with a knife or file.

The plates are now ready for pasting. They are placed on a smooth board and the holes in each plate filled with a stiff mixture of red lead and sulphuric acid (two parts water to one part acid). The paste must be pressed carefully into the holes with a flat stick. The plates are then laid aside to dry and harden. After they have thoroughly dried, they should be assembled as shown in the illustration. The plate in the center is

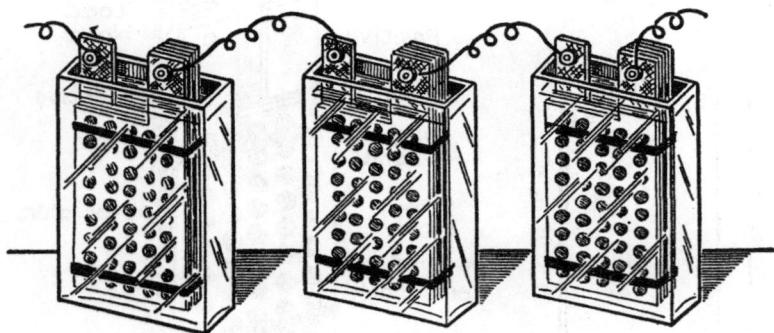

FIG. 78. — A Battery of Home-Made Storage Cells.

to be the positive. The two outside ones are negative. The wooden separators may be purchased at any automobile battery service station where repair work is done. The separators are easily cut to proper size with a sharp penknife. Each group of plates is then placed in a jar containing a mixture of sulphuric acid and water (four parts water to one part acid). On mixing the acid be certain to pour the acid into the water, stirring the mixture slowly at the same time, and not the water into the acid.

The plates are now ready for "forming." The binding-posts on the lugs of the plates may be secured from old dry cells. Connect the positive pole of a direct current source to the positive (center plate) of a cell and the negative pole to the negative plates. The cells of course may be placed in series and all three formed at the same time.

Allow the forming current to flow through the cells until the positive plate turns to a dark chocolate-brown color and the negatives to a slate-gray.

Lift the plates out, clean all sediment out of the jars and then replace them in the electrolyte. They will give considerable current. When this is exhausted, recharge immediately.

When the cells are fully charged, bubbles of gas will rise freely from both positive and negative plates. Recharging will only require about one-quarter of the time consumed in forming.

After the cells have been in use for some time, it is a good plan to lift out the plates and remove all sediment which has settled at the bottom of the jars.

A set of three such cells will have an E. M. F. of over six volts. Any number of cells may be connected in series in order to obtain a higher voltage.

Storage cells are usually rated in "ampere-hours." An ampere hour is the amount of current represented by one ampere flowing for one hour. A ten-ampere-hour storage cell will deliver:

> One ampere for ten hours
> Two amperes for five hours
> Five amperes for two hours
> Ten amperes for one hour.

In other words, the result obtained by multiplying the number of amperes by the time in hours is the ampere-hour capacity.

CHAPTER V

ELECTRO–MAGNETISM AND MAGNETIC INDUCTION

CONNECT two copper wires to a Voltaic cell and stretch one wire over a compass needle, holding it parallel to the needle and as near as possible without touching. Then bring the free ends of the wires together and observe that as the current flows, the needle is deflected and after a few moments back and forth comes to rest at an angle with the wire.

Next form a large loop of wire and place the needle within it. A greater deflection of the needle will occur this time when

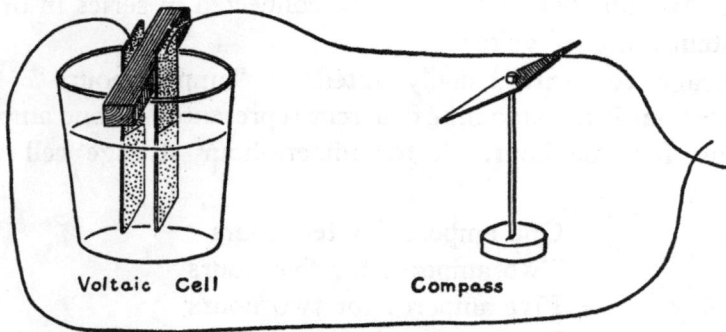

FIG. 79. — A Current of Electricity Flowing Through a Wire Will Deflect a Compass Needle.

a current flows through the wire. If a loop of several turns is formed, the deflection of the needle will be still greater.

As soon as the current is cut off the compass needle will return to its normal north and south position.

You have demonstrated a relationship between electricity and magnetism. These experiments were first performed by the Danish scientist Oersted, in 1819, and show that the region around a wire carrying a current of electricity has *magnetic* properties. In other words, an electric current produces magnetism.

Another interesting experiment showing the magnetic effect of a current of electricity may be performed by dipping a wire carry-

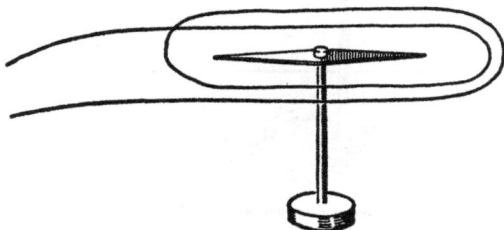

Fig. 80. — If a Loop of Wire Is Formed About a Compass Needle, the Deflection Will Be Greater.

ing a strong current of electricity into a pile of iron filings. A thick cluster of the filings will adhere to the wire. As soon as the current is cut off, the filings fall away from the wire, showing that the magnetic effect ceases with the current.

The current from a Leclanche type battery is not strong enough to make iron filings cling to the wire unless the filings are a fine dust. The stronger current from a small storage cell

Fig. 81. — Iron Filings Clustered on a Wire Carrying a Current of Electricity.

or a dichromate-of-potash cell is needed to make ordinary filings cling.

The region in the neighborhood of a wire carrying a current is a *field of force* through which lines of magnetism flow in

exactly the same way that they do in the neighborhood of a bar
or horseshoe magnet.

This is readily shown by punching a small hole in a piece of
cardboard, and passing a wire
carrying a strong current of electricity through the hole. If a few
iron filings are sifted on the cardboard and the latter jarred
slightly, the filings will arrange
themselves in circles with the wire
at the center. This magnetic
phantom shows the paths of the
lines of magnetic force around
the wire.

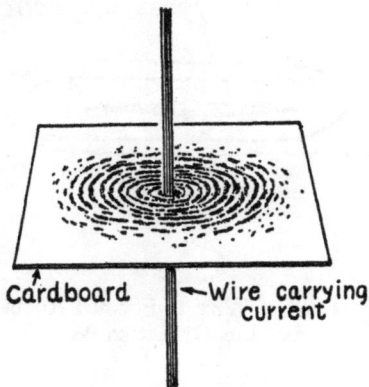

FIG. 82. — Magnetic Phantom
Formed About a Wire Carrying
a Current of Electricity.

By forming the wire into a coil
as shown in the illustration the
magnetic field generated is made
much stronger and the phantom is more plainly seen, for then
the combined effect of the wires is secured.

Roll up a small paper tube
about one-quarter of an inch
in diameter and three or four
inches long. Wind neatly on
the tube three or four layers
of magnet wire (any size
from No. 18 to No. 25).
Pass   an   electric   current
through it from two or three
cells of battery and test its
magnetic properties by bringing it near a compass needle.
It will be found that the coil

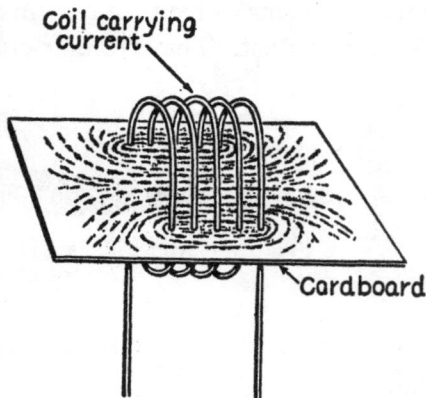

FIG. 83. — Magnetic Phantom Formed
About Several Turns of Wire.

possesses very marked magnetic properties, and will readily cause the needle to swing about, even though it is some distance from the coil.

If a large iron nail or a piece of iron rod is placed inside of the paper tube, the magnetic effect will be greatly increased. The presence of the iron bar inside of the coil of wire greatly increases the number of lines of magnetic force passing through the coil. Without the iron core, many of the lines

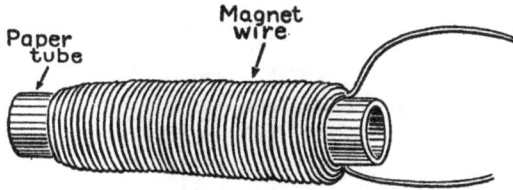

FIG. 84. — Paper Tube Wrapped with Wire for Experimental Purposes.

of magnetic force leak out at the sides of the coil, and but few extend from end to end. The effect of the iron core is not only to diminish the leakage of the lines of force, but also to add many more to those previously existing. Thus the magnetic strength of a coil is greatly increased by an iron core.

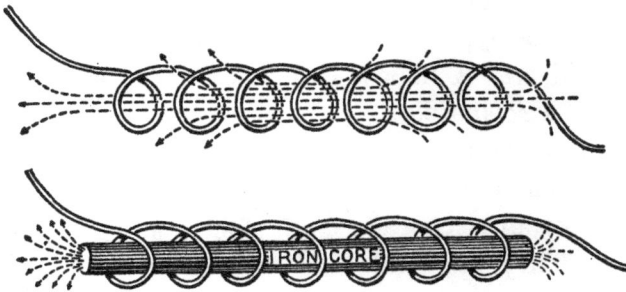

FIG. 85. — The Lines of Force "Leak" from a Coil of Wire, and Are Concentrated by an Iron Core.

A coil or wire wrapped around an iron core forms an *electro-magnet*.

If you wrap some insulated wire around an ordinary iron nail or a small iron bolt and connect it to one or two cells of battery it will become an electro-magnet and pick up bits of iron and

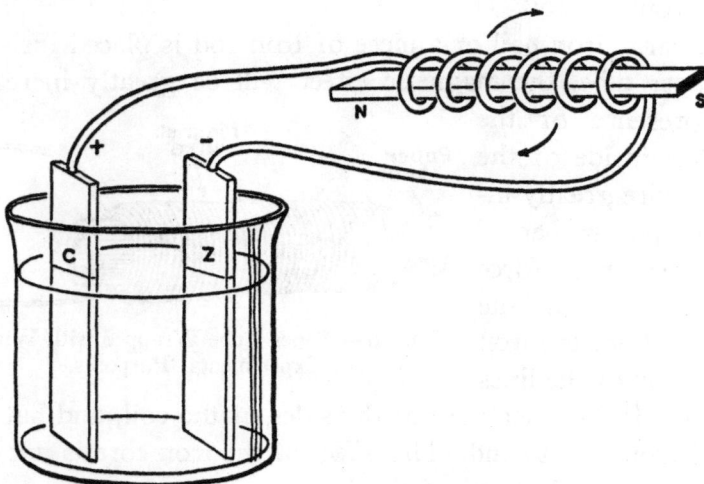

FIG. 86. — The Principle of an Electro-magnet

steel. If you wind the wire around a small paper tube into which a nail will slip easily, the coil will draw the nail in when the current is turned on. A hollow coil of this sort with a movable core is called a *solenoid*.

Electro-magnets and solenoids are an important part of almost all electrical machinery. They form an essential part of dynamos, motors, telephone receivers, telegraph relays, sounders, bells, radio speakers, voltmeters, ammeters, horns, and a host of other devices.

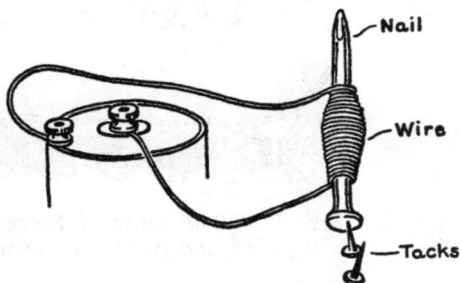

FIG. 87. — If you Wrap Some Insulated Wire Around an Ordinary Nail and Connect It to a Battery, It Will Become an Electro-magnet.

The form given to an electro-magnet depends upon the use to which it is to be put. A horseshoe form is common. This consists of two electro-magnets mounted on a yoke and connected so that the two free poles are north and south.

It is important to know how to make an efficient electro-magnet when designing or constructing electrical machinery. It is easier to explain how to make an efficient electro-magnet after the meaning of certain electrical terms such as volt, ampere, and ohm are understood. So you will find more about electro-magnets in Chapter IX.

MAGNETIC INDUCTION

In 1831, Michael Faraday discovered that if a magnet be suddenly plunged into a hollow coil of wire, a momentary current of electricity is generated in the coil. As long as the magnet remains motionless, it *induces* no current in the coil, but when it is *moved,* it sets up a current.

The source of electrical energy is the mechanical work done in moving the magnet. The medium which changes the mechanical energy into electricity is the magnetic field which we have already seen exists in the neighborhood of a magnet.

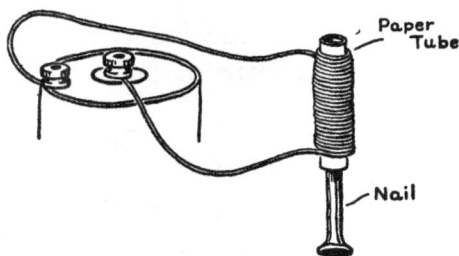

FIG. 88. — If You Wind Wire Around a Small Paper Tube into Which a Nail Will Slide Easily, the Coil Will Draw the Nail In When the Current is Turned On.

A current of electricity produced in a coil in such a manner is said to be an *induced* current and the phenomenon is that known as *magnetic induction*. Faraday's discovery was an important one because the principle of magnetic induction makes it possible to build dynamos, induction coils, telephone transformers, some forms of motors, magnetos and many other electrical devices. Without dynamos, by means of which electricity may be generated cheaply and in vast quantities, it would have been impossible to use electricity for light and power. The electricity produced by batteries is too expensive for such purposes.

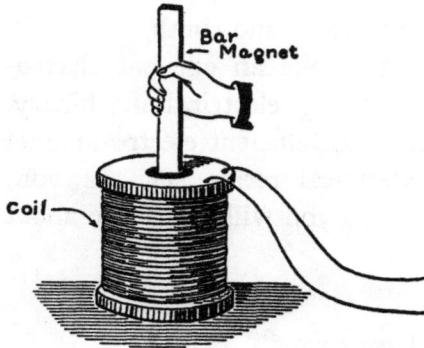

FIG. 89. — A Current of Electricity May be Induced by a Bar Magnet and a Coil.

A simple experiment in which electricity is produced by magnetic induction may be performed by winding a number of turns of fine insulated wire around the armature

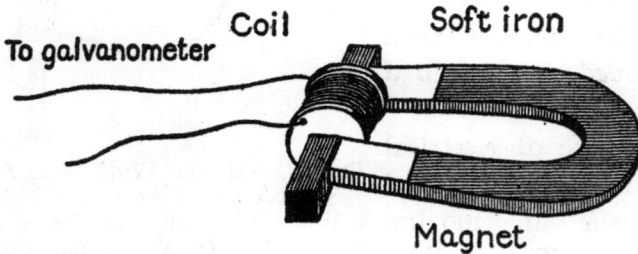

FIG. 90. — A Horseshoe Magnet and a Coil Arranged to Produce Electric Currents by Induction.

or keeper of a horseshoe magnet. The wire should be wound in the center of the keeper, leaving the ends of the iron free to come in contact with the poles of the permanent magnet.

Connect the ends of the coil to a sensitive galvanometer,* the ends of the keeper being in contact with the poles of the horseshoe magnet as shown in the illustration.

Keeping the magnet fixed, suddenly pull off the keeper. The galvanometer needle will swing showing a momentary current has been induced in the coil. Suddenly bring the coil up to the poles of the magnet; another momentary current, this time in the reverse direction will be indicated by the galvanometer. The needle will swing in the opposite direction this time.

It will be noticed that no current is produced when the coil and magnet are stationary. Current is generated only when the coil and magnet are approaching one another or moving apart suddenly.

This is because it is only there that the magnetic field passing through the coil is changing rapidly. The magnetic field is strongest near the magnet, and therefore if either the magnet or the coil of wire is moved, the strength of that part of the field which intersects the coil is changed. Induced currents can be generated only by changing magnetic field.

* See Chapter VIII.

Chapter VI

ELECTRICAL UNITS

THERE are certain terms used in electrical science to distinguish various properties and qualities of an electric current with which it is well for the young experimenter to acquaint himself.

The first thing required in order to make intelligent comparisons is a standard or unit of measure. The quart is the unit of *measure* commonly applied to liquids and is based upon the amount of space occupied by a certain volume. A quart of alcohol and a quart of milk occupy exactly the same space. The fact that they are different substances makes no difference. A quart of alcohol and a quart of milk do not weigh the same however.

The pound is a unit of weight which determines a certain amount of any substance by comparing the force, which gravity exerts in pulling that substance to the earth, with the same effect on another *standard* weight.

An electric current brings special problems when we attempt to compare or measure it. An electric current is invisible and weightless, and for these and other reasons cannot be measured by the quart or weighed by the pound. The only way that electricity can be measured is by means of some of the effects which it produces. Either the amount of magnetism, the heat or the chemical change which it will produce may be made the basis of a system of measurement.

The Ampere

The first method used to measure electric current was a chemical one.

If a current is passed through a solution of copper sulphate (blue vitriol) by means of two copper plates, copper will be dissolved from one plate and deposited on the other. If the current is furnished by a battery, the copper will be deposited on the plate connected to the zinc of the battery. If the current is allowed to flow for a short time and the two copper plates are then taken out and weighed, it will be found that one plate is distinctly heavier than the other.

The copper has been taken from one plate and deposited on the other by the electric current. A certain amount of current will always deposit the same amount of copper. By weighing the amount of copper deposited we have a method for measuring the amount of current which passed from one plate to the other.

The amount of electric current which will deposit 1.177 grams of copper in an hour is called an *ampere*. The word *ampere* is used to commemorate the memory of Andre M. Ampere, a French scientist who discovered many valuable things about the motion of electric currents. The ampere is the unit of electrical current measurement and implies quantity, volume or amount.

The chemical method of measuring current was at one time put to practical use in measuring the current used for electric lights and power. The first meters used in the first electrical power system (New York Edison Co.) consisted of a jar containing two copper plates immersed in a copper sulphate solution. The "meter" was connected so that all of the current used in a particular house or building would flow through it and cause copper to deposit on one plate. By weighing the plates every month, the power company could determine the amount of current consumed, and thereby determine the amount of the bill.

Nowadays, the electric meters make use of the magnetic effects of the current instead of the chemical as explained in one of the following chapters.

THE VOLT

For purposes of explanation an electric current may be likened to a stream of water flowing through a pipe.

If you let the water from a faucet run into a barrel and find that it will fill a barrel twice in one hour you have measured the volume of water flowing. It is at the rate of *two barrels per hour*. If you weigh the amount of copper deposited by an electric current in one hour and it is 2.354 grams you have measured the volume of current flowing. It is at the rate of *two amperes per hour*.

When you hold your thumb over the end of a water pipe or faucet from which water is flowing, the water will push against your thumb because of the *pressure* which the water exerts.

Just as a stream of water has both volume and pressure, so also does an electric current. Electric currents exert a pressure which is not called *pressure* in electrical parlance, but spoken of as *electromotive force* (abbreviated E. M. F.), potential, or voltage.

A stream of water must have pressure in order to move. Pressure enables it to pass through small openings and to overcome the resistance offered by a pipe. The pipe is a conductor which carries the water from place to place and corresponds to the conductors or wires which carry an electric current from place to place. A water pipe, due to the friction of the water against the pipe, offers resistance to the movement of the water.

Wires and other electrical conductors do not offer a perfectly free path to an electric current, but also possess a resistance. It is the pressure of the electric current which overcomes the resistance and pushes the current through the wire.

Water pressure may be measured in pounds but the pressure of an electric current is measured by a unit called the volt (after Volta). The volt is the unit of electrical pressure which will cause a current of one ampere to overcome a resistance in the circuit of one *ohm*.

THE OHM

The *ohm* * is the unit of electrical resistance. If we connect an ammeter to a battery, the ammeter will tell us how much current, *how many amperes* are flowing in the circuit. If we change the wire used in making the connections and use a smaller wire and perhaps a wire made of iron instead of copper we may find that only one-half as much current flows in the circuit as formerly. That is because the smaller iron wire offers more resistance to the current. *It has more ohms.* The resistance of a conductor depends upon its size, its material, its hardness, and its temperature.

The *standard* ohm, and by standard ohm is meant the ohm which the whole world uses, is the resistance offered to an electric current by a column of pure mercury having a section of one square millimeter and a length of 106.28 centimeters at a temperature of 0 degrees Centigrade.

The electrical pressure which will force sufficient current through such a column of mercury to deposit 1.177 grams of

---

* Named after Georg S. Ohm, the German physicist who discovered the laws of resistance.

copper in one hour is a volt, and in so doing has passed a current of one ampere through a resistance of one *ohm*.

Although the ohm, the ampere and the volt are different units of measure they bear a relation to each other which is explained by Ohm's Law.

Ohm's Law is a simple statement of facts which it is well for the boy electrician to understand thoroughly, for it might almost be said to be the basis of design of all electrical circuits and apparatus.

It is stated thus:

The strength of a current equals the voltage divided by the total resistance of the circuit.

In symbols, it is written

$$I = \frac{E}{R}$$

where I = the current in amperes

E = the electromotive force

R = the total resistance of all parts of the circuit.

By way of a simple example to show how Ohm's Law is used, we will suppose that a small telegraph sounder is connected to a battery and that the E. M. F. of the battery is *ten volts*. We will further suppose that the resistance of the sounder, connecting wires, and the battery itself is *five ohms*. Knowing these two facts, it is easy to find out how many amperes are flowing through the sounder by substituting these values in the equation as follows:

$$I = \frac{E}{R}$$

E = 10 volts and R = 5 ohms

$$I = \frac{10}{5} \text{ or 2 amperes.}$$

The Meaning of Milli and Kilo

In order to indicate fractions of very large values of electrical units it is customary to use the prefixes *milli, kilo,* etc., together with the words volt, ampere, etc.

Milli means one-thousandth; kilo means one thousand, and meg means one million. Micro means one-millionth.

$$\text{One millivolt} = \frac{1}{1000} \text{ of a volt}$$

$$\text{One kilovolt} = 1000 \text{ volts}$$

$$\text{One microvolt} = \frac{1}{1,000,000} \text{ of a volt}$$

$$\text{One milliampere} = \frac{1}{1000} \text{ of an ampere}$$

$$\text{One microampere} = \frac{1}{1,000,000} \text{ of an ampere}$$

$$\text{One megohm} = 1,000,000 \text{ ohms}$$

$$\text{One kilowatt} = 1000 \text{ watts}$$

The Watt

It is no doubt perfectly plain that a stream of water flowing through a pipe at a pressure of 100 lbs. per square inch is more powerful than an equal sized stream at 25 lbs. pressure.

Likewise a current of electricity represents more power at 100 volts potential than the same current would at 25 volts. Mechanical power or the ability to do a certain amount of work is measured by foot-pounds or horsepower.

The unit of electrical power is called the *watt.* A watt is the power represented by a current of one ampere flowing through a wire at a potential of one volt.

The number of watts flowing in a circuit is found by multiplying the voltage by the amperage. In the case of the circuit containing the sounder and battery used as an example to explain Ohm's Law the amount of *power* in the circuit is twenty watts (10 volts × 2 amperes).

Seven hundred and forty-six watts represent one electrical horsepower. One thousand watts are called a *kilowatt*.

THE COULOMB

So far, in discussing and explaining the various units with which an electric current or a circuit is measured we have not taken into consideration the element of time.

In order to explain the importance of time in measuring current values, we can again compare electricity with a stream of water.

If water is permitted to run out of a pipe into a tank until ten gallons has passed, it is not possible to tell at what *rate* the water is flowing until it is known how much *time* elapsed. Ten gallons per *second,* ten gallons per *minute* or ten gallons per *hour* would indicate the rate of flow.

One ampere flowing for one second is an electrical unit called the *coulomb* and is the means by which electrical rate of flow is expressed.

One ampere flowing for one hour is called an *ampere hour.* The number of ampere hours is found by multiplying the current in amperes by the time in hours.

The ampere hour is the means by which the capacity or ability of a battery to deliver electricity may be expressed. If a battery has a capacity of 10 ampere hours, this means that it will deliver one ampere for 10 hours (1 ampere × 10 hours = 10 ampere hours).

The same element of time enters into consideration in connection with the watt. One watt flowing for one hour is a *watt hour* and one kilowatt flowing for one hour is a *kilowatt hour*.

Chapter VII

WIRES AND ACCESSORIES

ELECTRIC currents are usually led from place to place at will by means of wires or cables. Cables are a group of wires bound together. There are a great many kinds of wires, each adapted to some special purpose.

Wires are usually covered with a non-conducting material called *insulation*. The insulation prevents the loss of electric current if the wires come into contact with other bodies or circuits.

The wires used in the interior of buildings are usually insulated with rubber, over which is placed a cotton braid to protect the rubber. Rubber is an excellent insulator but cannot be used as an insulator for all wires because it deteriorates with age.

"Weather-proof" wires, adapted to out-of-door service where they are exposed to the action of the elements, are insulated by heavy braids of cotton fiber and then impregnated with some compound, such as creosote and certain waxes.

Wires are made in a variety of insulations. Some may have only one insulating layer while others have a great many. Different substances are used as insulators to adapt the wire to some special purpose. Copper is usually the metal used to form the wire or conductor itself. Copper is a better conductor than any other metal except silver, the cost of which prohibits its use for such purposes. Aluminum is used as a current conductor, especially in transmission lines where it is an advantage to save weight in the spans between towers. Wires are also made up of special alloys or mixtures of metals purposely designed to possess resistance. These resistance wires will be explained later.

Wires may be solid, or made up of a number of small conductors so as to be flexible. The flexible wires used for connecting lamps and household devices are made up of a number of small wires twisted together and are called "cords."

In order that the conductors used for heavier currents may be rendered more flexible for handling they are made up of a number of wires or strands which are not insulated from one another but bound together with an insulation of rubber, tape, and braided cotton.

The telephone companies use both overhead and underground cables made up of a bundle of copper wires insulated from one another with paraffined paper and protected against mechanical injury by a lead sheath.

The wires which the young experimenter is most likely to use are known as magnet wires and are used for making electromagnets, coils and various windings. Magnet wires may be insulated with either silk, cotton, or enamel.

Silk-covered and cotton-covered wires may be obtained with either a single or a double covering.

Wires with a single covering of silk or enamel are used when it is desirable to save space, for these insulations are thinner than cotton or double-silk and consequently take up less space.

The size of a wire is indicated by its diameter, and in the United States is measured by the Brown and Sharpe gauge, usually indicated by the abbreviation "B. & S."

The tables show the various sizes of wire of the Brown and Sharpe gauge, and also several of their characteristics, such as weight, resistance, etc.

NUMBER, DIAMETER, WEIGHT, LENGTH, AND RESISTANCE OF COPPER WIRE

GAUGE. B. & S. No.	DIAMETER In Decimals of an Inch	SECTIONAL AREA In Circular Mils.	CAPACITY In Amp.	OHMS. Per 1,000 Feet.	OHMS. Per Mile.	OHMS. Per Pound.	FEET. Per Pound.	FEET. Per Ohm.	POUNDS. Per 1,000 Feet.	POUNDS. Per Ohm.
0000	.460	211600.	312.	.04906	.25903	.000077	1.56122	20497.7	640.51	12987.
000	.40964	167805.	262.	.06186	.32664	.00012	1.9687	16255.27	507.95	8333.
00	.3648	133079.	220.	.07801	.41187	.00019	2.4824	12891.37	402.83	5203.
0	.32486	105534.	185.	.09831	.51909	.00031	3.1303	10223.08	319.45	3225.
1	.2893	83694.	156.	.12404	.65490	.00049	3.94714	8107.49	253.34	2041.
2	.25763	66373.	131.	.1563	.8258	.00078	4.97722	6429.58	200.91	1282.
3	.22942	52634.	110.	.19723	1.0414	.00125	6.2765	5098.61	159.32	800.
4	.20431	41743.	92.3	.24869	1.313	.00198	7.9141	4043.6	126.35	505.
5	.18194	33102.	77.6	.31361	1.655	.00314	9.97983	3206.61	100.20	318.
6	.16202	26251.	65.2	.39546	2.088	.00499	12.5847	2542.89	79.462	200.
7	.14428	20817.	54.8	.49871	2.633	.00792	15.8696	2015.51	63.013	126.
8	.12849	16510.	46.1	.6529	3.3	.0125	20.0007	1599.3	49.976	80.
9	.11443	13094.	38.7	.7892	4.1	.0197	25.229	1268.44	39.636	50.
10	.10189	10382.	32.5	.8441	4.4	.0270	31.8212	1055.66	31.426	37.
11	.090742	8234.	27.3	1.254	6.4	.0501	40.1202	797.649	24.924	20.
12	.080808	6530.	23.	1.580	8.3	.079	50.5906	632.555	19.766	12.65
13	.071961	5178.	19.3	1.995	10.4	.127	63.7948	501.63	15.674	7.87
14	.064084	4107.	16.2	2.504	13.2	.200	80.4415	397.822	12.435	5.00
15	.057068	3257.	13.6	3.172	16.7	.320	101.4365	315.482	9.859	3.18
16	.05082	2583.	11.5	4.001	23.	.512	127.12	250.184	7.819	1.95
17	.045257	2048.	9.6	5.04	26.	.811	161.29	198.409	6.199	1.23
18	.040303	1624.	8.1	6.36	33.	1.29	203.374	157.35	4.916	.775

Gauge. B. & S. No.	Diameter. In Decimals of an Inch	Sectional Area. In Circular Mils.	Capacity. In Amp.	Ohms. Per 1,000 Feet.	Ohms. Per Mile.	Ohms. Per Pound.	Feet. Per Pound.	Feet. Per Ohm.	Pounds. Per 1,000 Feet.	Pounds. Per Ohm.
19	.03589	1288.	....	8.25	43.	2.11	256.468	124.777	3.899	.473
20	.03196I	1021.	....	10.12	53.	3.27	323.399	98.9533	3.094	.305
21	.028462	810.	....	12.76	68.	5.20	407.815	78.473	2.452	.192
22	.025347	642.	....	16.25	85.	8.35	514.193	62.236	1.945	.119
23	.02257I	509.	....	20.30	108.	13.3	648.452	49.3504	1.542	.075
24	.0201	404.	....	25.60	135.	20.9	817.688	39.1365	1.223	.047
25	.0179	320.	....	32.2	170.	38.2	1031.038	31.0381	.9699	.030
26	.01594	254.	....	40.7	214.	52.9	1300.180	24.6131	.7692	.0187
27	.014195	201.	....	51.3	270.	84.2	1639.49	19.5191	.6099	.0112
28	.012641	159.8	....	64.8	343.	134.	2067.364	15.4793	.4837	.0073
29	.011257	126.7	....	81.6	432.	213.	2606.959	12.2854	.3835	.0047
30	.010025	100.5	....	103.	538.	338.	3287.084	9.7355	.3002	.0029
31	.008928	79.7	....	130.	685.	539.	4414.49	7.72143	.2413	.0018
32	.00795	63.	....	164.	865.	856.	5226.915	6.12243	.1913	.0011
33	.00708	50.1	....	206.	1033.	1357.	6590.41	4.85575	.1517	.00076
34	.006304	39.74	....	260.	1389.	2166.	8312.8	3.84966	.1204	.00046
35	.005614	31.5	....	328.	1820.	3521.	10481.77	3.05305	.0956	.00028
36	.005	25.	....	414.	2200.	5469.	13214.16	2.42I7	.0757	.0018
37	.004453	19.8	....	523.	2765.	8742.	16659.97	1.92086	.06803	.00014
38	.003965	15.72	....	660.	3486.	13772.	21013.25	1.52292	.04758	.00007
39	.003531	12.47	....	832.	4395.	21896.	26496.237	1.20777	.03755	.00064
40	.003144	9.88	....	1049.	5542.	34823.	33420.63	0.97984	.02992	.000024

INSULATORS

The covering placed over wires is not the only precaution taken to insulate them. They are sometimes supported on glass or porcelain *insulators*.

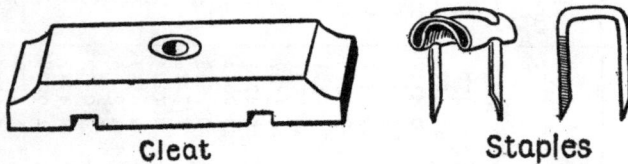

Cleat   Staples

FIG. 91.— Staples and Wooden Cleat Used for Running Low Voltage Wires.

Telegraph and telephone lines are usually supported on glass knobs. Power lines are insulated with porcelain.

RESISTANCE WIRES

The *longer* wires are, and the thinner they are, the more will they resist. But there is another way of giving a wire resistance besides making it long and thin. By making it of a suitable alloy such as a mixture of nickel and chromium (Nichrome) or of

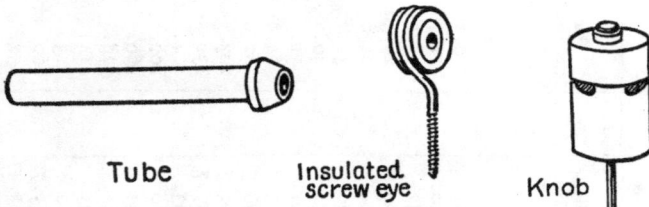

Tube   Insulated   Knob
            screw eye

FIG. 92. — Porcelain Insulators to Support Electric Light Wires.

copper, manganese and nickel a great deal of resistance can be obtained from a short piece of wire.

Of what use is resistance? We shall see.

An electric current may do work of various kinds, chemical, mechanical, magnetic, and thermal. In every instance, however, where a current does work, that work is done by the expenditure of part of the energy that is being supplied to the circuit. Part of the energy supplied to a circuit is expended in overcoming the resistance. When electricity in motion is opposed by resistance the energy which it uses up in overcoming the resistance is changed into heat. Heat appears wherever there is resistance to a current.

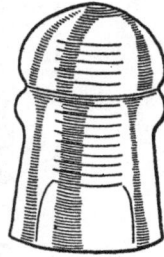

Insulator Pin

FIG. 93. — Glass Insulator and Pin Used to Support Telegraph and Telephone Poles.

Electric ovens, flatirons, soldering irons, toasters and many other similar devices contain a resistance element usually of Nichrome wherein an electric current is changed into heat.

BINDING-POSTS

Binding-posts are a simple device for clamping a wire so as to make a quick and convenient electrical connection.

FIG. 94. — Types of Binding-Posts.

Binding-posts may be either made or purchased. Those which are purchased are the best and usually add to the appearance of home-made electrical equipment.

The knurled thumb nuts obtained from old dry cells make

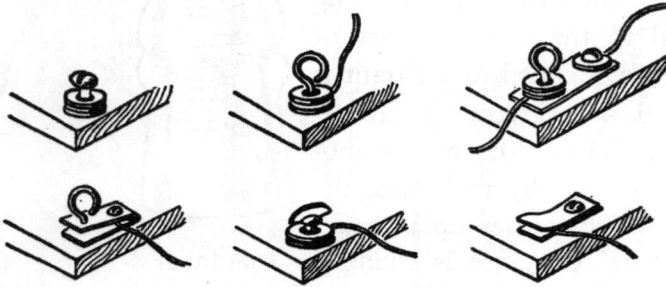

FIG. 95. — Home-Made Binding-Posts.

convenient binding-posts when completed with a hexagonal nut and screw.

The home-made binding-posts shown in the sketch should be made out of *brass*. Do not use iron screw-eyes, washers, etc., because they do not provide a low resistance contact.

Thumbnut

Hex nut

8-32 Screw

FIG. 96. — Binding-Post Made from Thumbnut Removed from a Dry Cell.

SWITCHES AND CUT-OUTS

Switches and cut-outs are used for turning a current off and on. Several simple home-made switches are illustrated.

The first one shown (A) has one contact, formed by driving a brass escutcheon pin (a round-headed brass nail) through a small strip of copper or brass.

The movable lever is a strip of copper or brass, rolled up to form a handle at one end. The other end is pivoted with a brass screw. The brass screw passes through a small strip of copper or brass having a binding-post mounted on the end. A small copper washer should be placed between the lever and the copper strip to permit the lever to swing more easily.

A somewhat similar switch is shown by B in the same

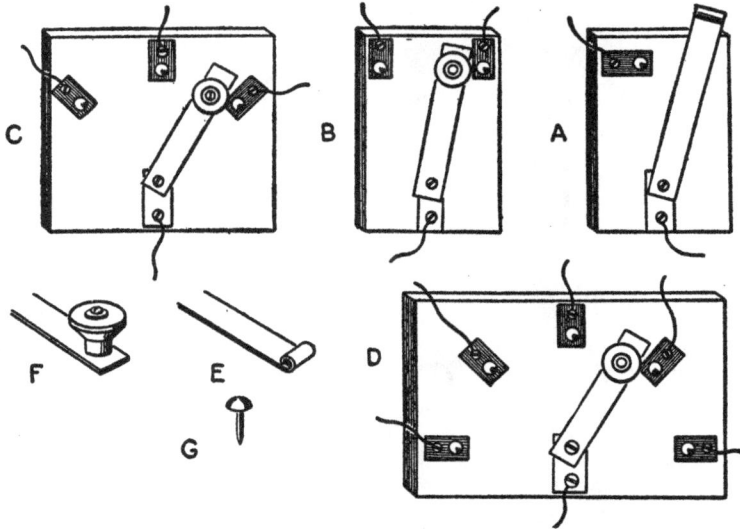

FIG. 97. — Simple Switches.

A. Single-Point Switch.
B. Two-Point Switch.
C. Three-Point Switch.
D. Five-Point Switch.
E. Lever with End Rolled up to form Handle.
F. Lever with Handle Made from Part of a Spool.

illustration, only in this case a handle made from half a spool is used, instead of rolling up the end of the arm.

The other sketches (C and D) show how the same method of construction may be utilized to make switches having more than one "point" or contact.

No dimensions have been given for constructing these switches, because it is doubtless easier for the young experimenter to use materials which he may have at hand and construct a switch of his own proportions. Bevel the under edges of the lever slightly with a file, so that it will slip over the head of the brass tack more easily.

The knife-switches shown in Fig. 98 will make better con-

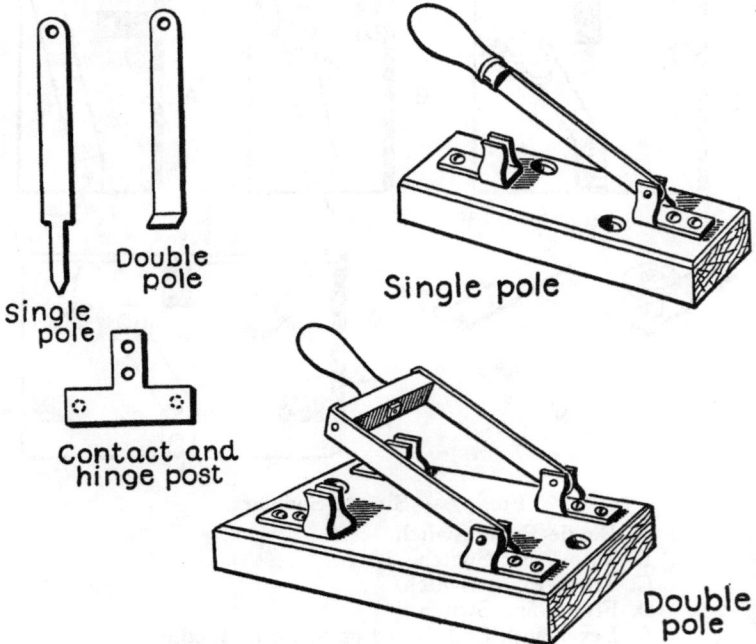

Single pole

Double pole

Single pole

Contact and hinge post

Double pole

FIG. 98. — Knife Switches.

tacts and carry heavier currents than those just described. They are the type used on power switchboards.

The base for a home-made knife-switch may be wood, but preferably should be some insulating material which will not absorb water such as bakelite, masonite, etc.

The details of the metal parts are shown in Fig. 98. The levers, contacts and hinge posts should be cut from sheet-brass or sheet-copper and bent into shape. The wooden handle of the single-pole switch is driven onto the metal tongue.

The double-pole switch is almost a duplicate of the single-pole type, but has two sets of levers and contacts. The ends of the blades to which the handle is fastened are turned over at right angles. The hardwood cross-bar is fastened between the ends with two small wooden screws. The handle is attached to the center of the cross-bar.

After the switch is assembled, bend the various parts until they "line up," that is, are in proper position in respect to each other, so that the blades will drop into the contacts without bringing pressure to bear on either one side or the other of the handle in order to force the blades into line.

FUSES

Fuses are used to protect electrical instruments and circuits from damage due to too much current.

A fuse is usually a short piece of lead or alloy which melts at a low temperature. It is placed in the circuit which it is to protect. If too much current flows the fuse will become hot and melt. Because of its low melting point, it melts before the other parts of the circuit become overheated. Thus the circuit is broken and the current shut off until the cause which occasioned the surplus current to flow can be ascertained.

Fuses are rated according to the amount of current required to "blow" them out and are called 1, 2, 3, 5, 10, or 20 ampere fuses, as the case may be.

The experimenter may easily make simple fuses to protect his batteries, etc., from short circuits.

The simplest possible fuse consists merely of a small piece of lead wire or a narrow strip of thick lead or tinfoil held between two binding-posts mounted upon a wooden block.

A small strip of tinfoil which will be "blown out" by a small amount of current may be supported on a piece of mica. A strip of thin sheet-copper is bent around the ends of the mica strip. Such a fuse is held in a mounting as shown by D. The contacts are made from sheet-copper or brass. They should spring together tightly, so as to make perfect contact with the copper ends of the mica strip.

Lightning-Arrestors

Lightning-arrestors are used to protect wires which enter buildings from outdoors, especially telegraph or telephone wires. The protection afforded by an ordinary lightning arrestor is not sufficient to ward off a bolt of lightning should it happen to directly strike the wire. The static charge induced by lightning striking in the neighborhood can however be safely dissipated into the ground.

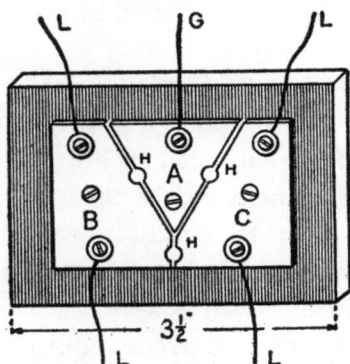

FIG. 99. — Home-Made Lightning-Arrestor.

Lightning-arrestors may be constructed in different ways but the only type for which the average experimenter will have any use is a simple "ground switch."

It consists of three pieces of sheet-brass about one-sixteenth of an inch thick, shaped as shown by A, B and C in Fig. 99. The metal pieces are mounted on a wooden block so as to leave a narrow space about one thirty-second of an inch separating them. The two outside pieces are each fitted with two binding-posts or terminal screws and the center triangular-shaped piece is fitted with one. A hole about one-eighth of an inch in diameter is bored between each of the metal pieces.

Make a tapered brass pin which can be placed tightly in the holes and will make contact with the metal pieces.

The two outside line wires of the telegraph or telephone circuit are connected to the outside metal pieces C and B. A is connected to the earth or ground.

In case of a lightning storm, if the wires become charged, the small space between the metal plates will permit the static charge to leak or jump across to A and pass harmlessly into the ground. If more complete protection is desired, it is merely necessary to insert the plug into one of the holes, and thus ground either wire or short circuit both of them.

Chapter VIII

ELECTRICAL MEASURING INSTRUMENTS

Galvanoscopes and Galvanometers

In the first part of Chapter V it was explained that several turns of wire surrounding a compass-needle would cause the needle to move if a current of electricity were sent through the coil.

Such an arrangement is called a galvanoscope and may be used for detecting feeble currents. A galvanoscope becomes a *galvanometer* by providing it with a scale so that the movement or deflection of the needle may be measured.

A galvanometer is really, in principle, an ammeter the scale of which has not been calibrated to read in amperes. An ammeter is an instrument designed to measure volume of current.

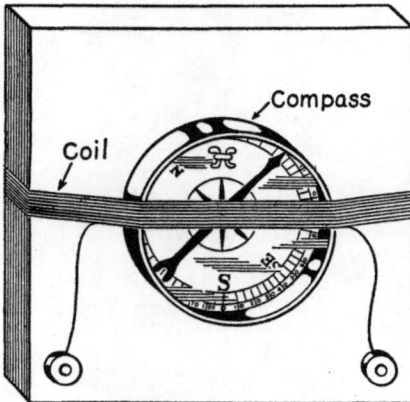

Fig. 100. — Simple Compass Galvanoscope.

A Galvanoscope

A simple galvanoscope may be made by winding fifty turns of No. 36 B. & S. gauge magnet wire around an ordinary pocket compass. The compass may be set on a block of wood and the terminals of the coil provided with binding-posts so that connections are easily made.

Another variety of the same instrument can be made by winding about twenty-five turns of No. 30 B. & S. gauge magnet wire around the lower end of a glass tumbler. After slipping the coil from the glass, tie the wire with thread in several places so that it will not unwind. Press two sides of the coil together

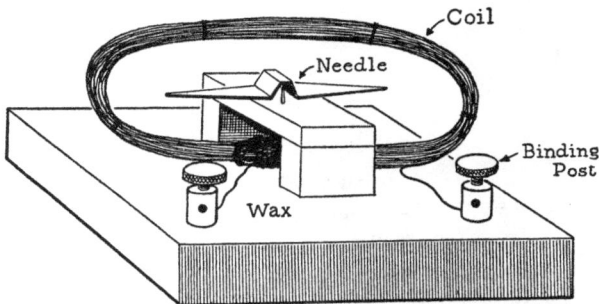

FIG. 101. — A Home-Made Galvanoscope.

so as to make it an elliptical shape and then attach it to a block of wood with sealing wax.

Make a little wooden bridge as shown in the illustration and mount a compass needle on it in the center. The compass needle may be made out of a piece of spring-steel in the manner described in Chapter I.

Mount two binding-posts at the corners of the block and connect the terminals of the coil to them. Turn the block so that the needle points North and South and parallel to the coil of wire. If a battery is connected to the binding-posts, the needle will turn around until it is almost parallel with the axis of the coil.

An Astatic Galvanoscope

An astatic galvanoscope is one having two needles, arranged so that their unlike poles are opposite. The word "astatic," used in connection with a magnetic needle, means having no directive magnetic tendency. If the needles of an astatic pair are separated and pivoted individually, they will each point to North and South in the ordinary manner. But when connected together with the poles arranged in opposite directions they neutralize each other. An astatic needle requires but very little current in order to turn it either one way or the other, and for this reason an astatic galvanoscope is usually very sensitive.

An instrument of this sort can be made by winding about fifty turns of Nos. 30 to 36 B. & S. gauge magnet wire into a coil around a glass tumbler. After removing the coil from the glass, shape it into the form of an ellipse and fasten it to a small base-board. Separate the strands of wire at the top of the coil so that they are divided into two groups.

Make a standard in the shape of an inverted U out of thin wooden strips and fasten it to the block.

The needles are ordinary sewing-needles which have been magnetized and shoved through a small carrier-bar, made from a strip of cardboard, with their poles opposite one another. They are held in place in the cardboard strip by a small drop of sealing-wax.

A small hole is punched in the top of the carrier, through which to pass the end of a thread. The upper end of the thread

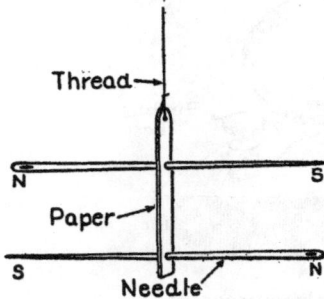

FIG. 102. — Astatic Needles.

passes through a hole in the bridge and is tied to a small screw-eye.

The carrier bar is passed through the space where the coil is split at the top. The lower needle should hang in the center of the coil. The upper needle should be above and outside the coil. The terminals of the coil are connected to two binding-posts mounted on the base-block.

Since this galvano-scope is fitted with astatic needles, the instrument does not have to be turned so that the needles and coil are in a north and south position. A very slight current of electricity passing through the coil will swing the needles.

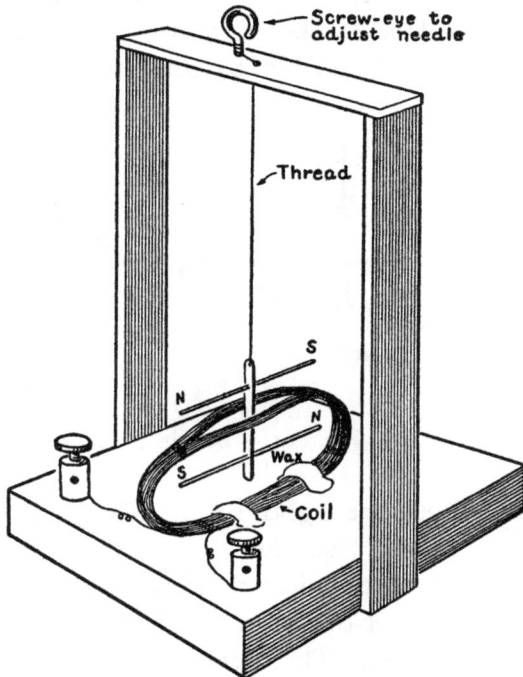

FIG. 103. — Astatic Galvanoscope.

AN ASTATIC GALVANOMETER

A sensitive astatic galvanometer for detecting exceedingly weak currents and for use in connection with a "Wheatstone bridge" for measuring resistance, as described farther on, will form a valuable addition to the laboratory of the boy electrician.

Make two small wooden bobbins from cigar box wood and

according to the shape and dimensions shown in the illustration. Do not use any nails in the construction of the bobbins. They should be fastened together with glue.

Wind each bobbin in the same direction with No. 36 B. & S. gauge silk or enamel-covered magnet wire, leaving about six inches free at the ends for connection to the binding-posts. Fasten each bobbin to the baseboard, parallel to each other

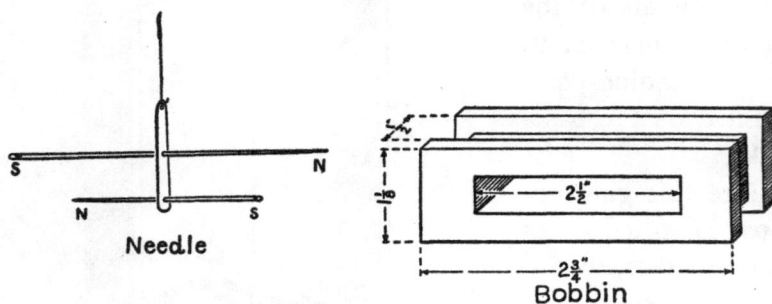

FIG. 104. — Bobbin for Astatic Galvanometer.

and about one-eighth of an inch apart. Do not nail the bobbins in position. Use glue.

Connect the coils wound on the bobbins so that the end of the outside layer of the first coil is connected to the inside layer of the other coil. The current will then pass through the windings in the same continuous direction, exactly as though the bobbin were one continuous spool.

Magnetize two small sewing-needles and mount them in a paper stirrup made from a strip of strong paper. The north pole of one needle should be on the same side of the stirrup as the south pole of the other. They are fastened in place with a small drop of melted sealing-wax.

Cut out a cardboard disk and divide it into degrees as shown in the illustration. Glue the disk to the top of the bobbins. A small slot should be cut in the disk so that it will pass the lower needle. The needles are suspended from a wooden arm fastened to the top of a wooden post glued to the base. A fine fiber for

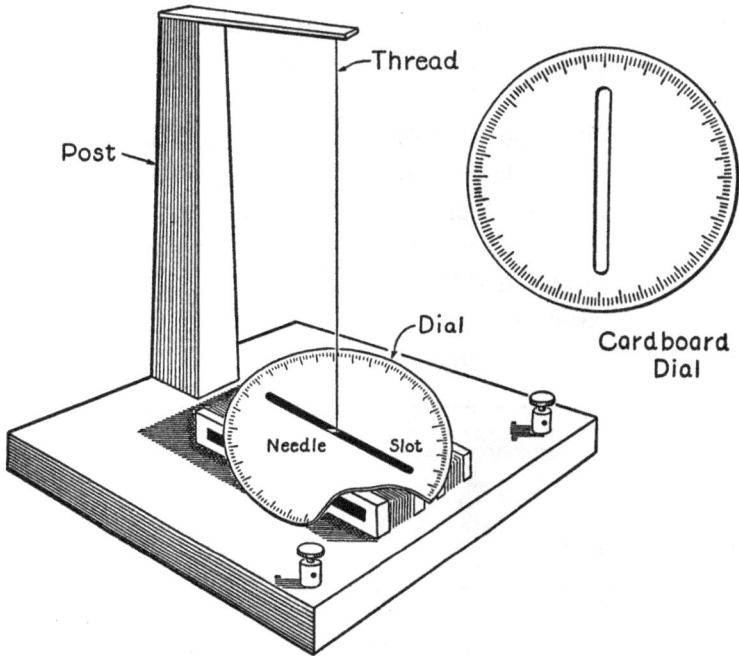

FIG. 105. — Completed Astatic Galvanometer.

suspending the needles may be secured by unraveling a piece of embroidery silk. The fiber should be as fine as possible; the finer it is, the more sensitive the instrument will be.

The upper end of the fiber is tied to a small hook in the end of the arm. The wire hook may be turned so that the needles can be brought to zero on the scale. Zero should lie on a line parallel to the two coils. The lower needle should swing inside of the two coils, and the upper needle above the disk.

How to Make a Wheatstone Bridge

The boy electrician will have many occasions when it is desirable to know the resistance of some of his electrical apparatus. Telephone receivers, telegraph relays and other electrical instruments are rated according to their resistance in ohms.

The simplest method of measuring resistance is by means of the device known as the Wheatstone bridge. This instrument is very simple but at the same time is remarkably accurate. A Wheatstone bridge is shown in Fig. 106. The measurement of resistance is accomplished by comparing the unknown resistance with a known resistance.

The base of the instrument is a piece of well-seasoned wood,

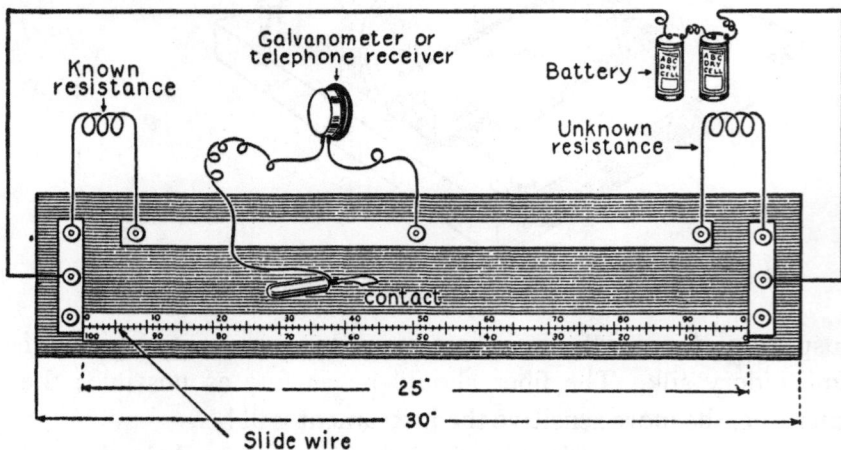

FIG. 106. — Wheatstone Bridge.

thirty inches long, six inches wide, and three-quarters of an inch thick.

Secure a long strip of No. 18 B. & S. gauge sheet-copper or brass one inch wide. Cut it into three pieces, making two of the pieces three inches long, and the other twenty-three and one-half inches long.

Mount the strips on the wooden base, as shown in the illustration, being careful to make the distance between the inside edges of the end strips exactly twenty-five inches. The strips should be fastened to the base with round-headed brass screws. Mount two binding-posts on each of the short strips in the positions shown in the illustration, and three on the long strip.

Then make a paper scale twenty-five inches long, and divide it into one hundred equal divisions (one-quarter of an inch apart). Mark every fifth division with a slightly longer line and every tenth division with a double length line.

Starting at one end, number every tenth division. Then start at the other end and number them back, so that the scale reads 0, 10, 20, 30, 40, 50, 60, 70, 80, 90, 100 from right to left at the top and 0, 10, 20, 30, 40, 50, 60, 70, 80, 90, 100 from left to right at the bottom.

Solder a piece of No. 30 B. & S. gauge German-silver wire to one of the short copper strips opposite the end of the scale.

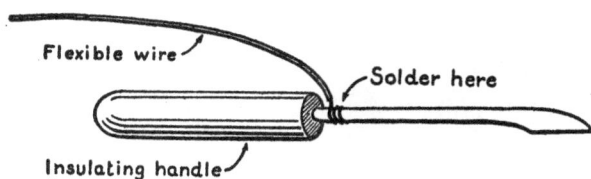

Fig. 107. — Knife-Contact.

Stretch the wire tightly across the scale and solder it to the strip at the other end.

Make a knife-contact by flattening a piece of heavy copper wire as shown in Fig. 107. Solder a piece of flexible insulated wire such as "lamp cord" to the other end. Fit the contact with a small wooden handle made from a piece of dowel.

The instrument is now ready to use.

In order to make resistance measurements with a Wheatstone bridge, it is necessary to have a set of resistances of known value. The resistance of any unknown circuit or piece of apparatus is then found by using the bridge to compare it with one of the known coils. The Wheatstone bridge is a sort of "balancing" device.

Using it may be compared to weighing sugar or some other substance on a pair of scales. The sugar is measured or weighed by balancing it on the scales with a known weight of one, two, five or ten pounds, as the case may be. The Wheatstone bridge might be called a pair of "electrical scales" for weighing resistance by comparing an unknown coil with one which we know has a certain value.

The next step is to make up some standards or resistance coils of known value. The coils are made of No. 32 B. & S. gauge magnet wire. It will make no difference whether the insulation is silk, cotton or enamel. Cut the wire into the following lengths, laying it out straight on the floor but using care not to pull it hard enough to stretch it.

Resistance	Length of Wire
½ ohm coil	3 feet ½ inch
1 ohm coil	6 feet 1¼ inches
2 ohm coil	12 feet 2½ inches
5 ohm coil	30 feet 6¼ inches
10 ohm coil	61 feet
20 ohm coil	122 feet
30 ohm coil	183 feet
50 ohm coil	305 feet
100 ohm coil	610 feet

These lengths of wire are then wrapped on spools in the following manner. The winding is known as the non-inductive method, because the windings do not generate a magnetic field. If they produced a magnetic field, they might disturb the needle of the galvanometer used in connection with the bridge.

Each length of wire should be doubled *exactly* in the middle, then wrapped on the spools like a single wire, the two ends being left free for sol-

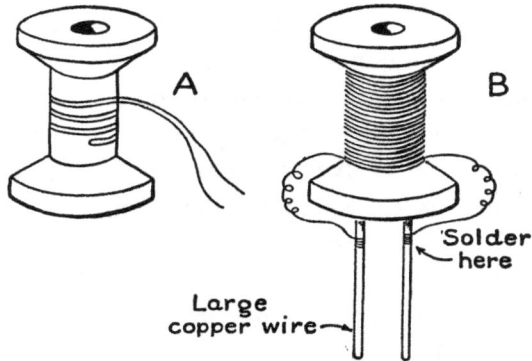

Fig. 108. — Resistance-Coil. *A*. Shows How the Wire is Doubled and Wound on the Spool. *B*. The Completed Coil.

dering to the terminals as shown in the illustration. The spools may be the ordinary reels upon which cotton and silk thread are wrapped. The spool for the 100 ohm coil will however need to be larger than the others.

The terminals of the spools are pieces of stout copper wire (No. 12 or No. 14 B. & S. gauge). Two pieces of wire about three inches long are driven into holes bored in the end of each spool. A small drop of solder is used to permanently secure the ends of the coil to the heavy wire terminals. The spools are then dipped in a pan of melted paraffin.

The spools should be marked ½, 1, 2, 5, 20, 30, 50 and 100 according to their resistance. The resistance of each spool will not be exactly the value given in the table but will be close enough so that the "standards" can be used for ordinary measurements.

How to Use the Wheatstone Bridge for Measuring Resistance

The instrument is connected as shown in the diagram.

The unknown resistance or device to be measured is connected across the gap at B. One of the standard coils estimated to be nearly the same number of ohms as the unknown resistance is connected across the gap at A. A sensitive galvanometer or a telephone receiver and two cells of battery are connected to the bridge as shown in the diagram.

If a telephone receiver is used, place it to your ear. If a galvanometer is used instead of a telephone receiver, watch the needle carefully. Then move the knife-contact over the scale along the German-silver "slide-wire" until a point is reached where there is no deflection of the needle or a minimum of sound in the receiver when the knife-contact is tapped against the slide-wire.

If this point lies very far on one side or the other of the center point on the scale, substitute the next higher or lower "standard" resistance spool until the point falls as near as possible to the center of the scale.

When this point is found, note the reading on the scale carefully. Now comes what may seem at first the hardest part. But almost all my readers have no doubt progressed far enough in arithmetic to be able to carry out the following simple calculation in proportion which must be made in order to find out the resistance of the unknown coil.

The value or number of ohms in the unknown resistance, connected to B, bears the same ratio to the known coil, at A, that the number of divisions between the point found on the slide-wire which you noted and the right hand end of the scale bears to the number of divisions between that point and the left hand end of the scale.

In order to make this explanation clearer, we will suppose that a 5 ohm coil was used at A and the needle of the galvanometer showed no deflection when the knife-contact rested on the 60th division from the left hand end, or the 40th from the right.

Then to find the value of the unknown resistance at B it is simply necessary to multiply it by 40 and divide by 60. The answer will be the resistance at B in ohms.

The calculation in this case would be as follows:

$$5 \times 40 = 200$$
$$200 \div 60 = 3.33 \text{ ohms}$$

3.33 ohms is the resistance at **B**.

This explanation may seem long and complex but if you study it carefully you will find it actually to be simple. When you master it, you will be able to make resistance measurements which will add greatly to the interest and value of your experiments.

VOLTMETERS AND AMMETERS

In addition to measuring the resistance of a circuit it is frequently necessary to measure the pressure and volume of a current.

An instrument designed to measure electromotive force (electrical pressure) is called a *voltmeter*. An instrument designed to measure volume of current is called an *ammeter*.

Some types of voltmeters and ammeters are more carefully made than a fine watch while others, for example the little meters used on the instrument panel of automobiles, are comparatively crudely constructed.

The little meters described in this chapter are simple and inexpensive but quite sensitive. Two different meters are described. Both operate on exactly the same principle but one is more elaborate than the other.

How to Make a Simple Voltmeter and Ammeter

The base is a wooden block, five inches long, two and one-half inches wide and one-half inch thick. In its center, cut a slot three-eighths of an inch wide and one and one-half inches long, with the slot running lengthwise of the board. Along each side of the slot glue two small wooden blocks one and one-half inches long, one-quarter of an inch thick, and one-half of an inch high. Glue a strip of wood, two and one-half inches long, three-quarters of an inch wide and one-eighth inch thick to the top as shown by D in the illustration.

These parts form a bobbin, integral with the base. If the

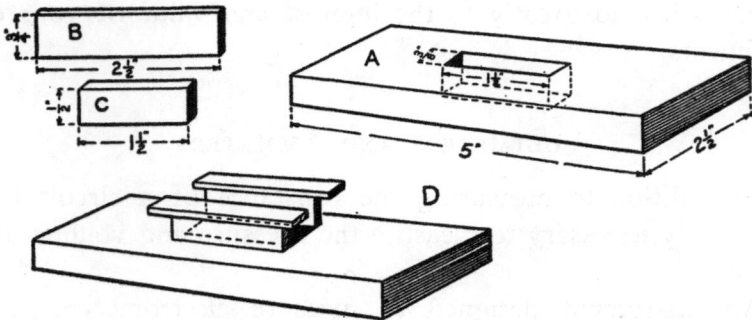

Fig. 109. — *A*, Base, Showing Slot. *B* and *C*, Sides and Top of the Bobbin. *D*, Base and Bobbin in Position.

meter is to be a voltmeter, wind on 200 feet of No. 36 B. & S. gauge enameled wire; if an ammeter, wind the coil full of No. 16 B. & S. gauge enameled wire.

The needle is a piece of watch spring, about one and one-quarter inches long, and one-eighth of an inch wide. Straighten the piece of spring and then heat it at its center with a small alcohol flame so as to anneal it slightly at that point.

The needle is mounted on a small steel shaft made of a piece of ordinary sewing needle one-half inch long. It must have sharp points ground on both ends with a small emery wheel or Carborundum stone. Bore a small hole through the center of the spring just large enough to receive the sewing needle shaft.

If the piece of watch spring lost its temper or hardness other than at its center it must be hardened again so that it will retain magnetism. It may be hardened by heating red-hot and plunging suddenly into cold water. It is then magnetized by winding

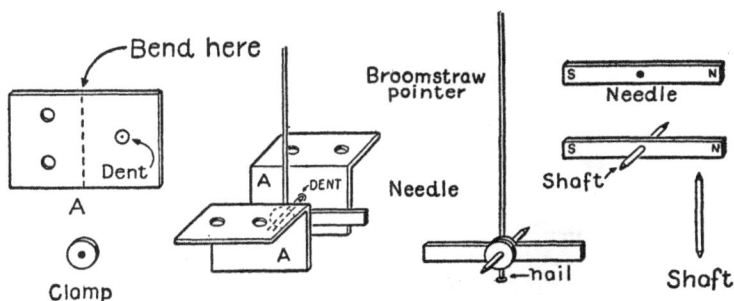

FIG. 110. — Arrangement of the Needle and Pointer.

ten or twelve turns of magnet wire around one end and connecting it with a battery for a moment.

Insert the shaft in the hole in the magnetic needle and fasten it in the center by two small wooden washers or dishes which fit tightly on the shaft and are cemented in place.

The pointer is a piece of straight broom-straw, about three inches long.

The pointer is a piece of straight broom-straw, fastened to the top of the magnetic needle with cement. It should be in a position at right angles to the needle. A small wire nail should be cemented on the opposite side of the needle from the broom-straw to serve as a counterweight and keep the pointer in a vertical position.

The needle is supported by two pieces of thin sheet-brass, one inch long and one-half inch wide. Bend each strip to form a right angle in the center and one-quarter of an inch from one end make a small dent by means of a center-punch and a hammer.

The strips, which we will now call the bearings, are slipped down into the slot in the bobbin with the dents inside of the

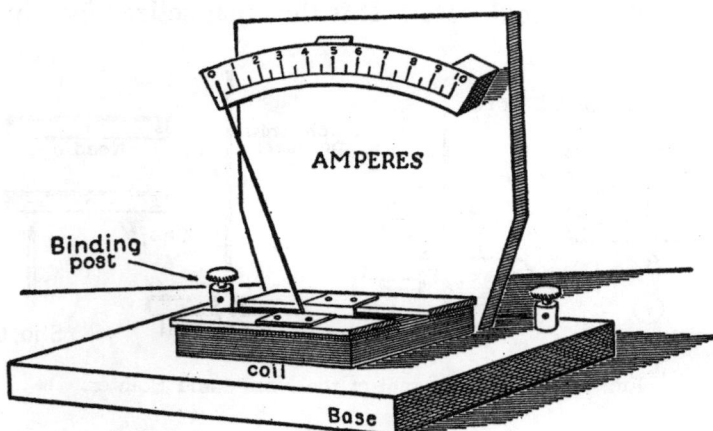

FIG. 111. — The Completed Meter.

coil and exactly opposite one another. After they are properly located they should be permanently fastened with cement or two very small screws.

The magnetized needle with its sharp pointed shaft, pointer and counterweight, should slip inside the bobbin with the ends of the shaft resting in the dents in the bearing strips and there swing freely. It may require a little careful filing and bending to accomplish this, but the work should be done patiently, because the proper working of the meter will depend upon having the needle swing freely and easily in its bearings.

Fasten an upright board, four inches wide and one-quarter of an inch thick, to the base-board, back of the bobbin.

Attach a semi-circular piece of thick cardboard to the upright by means of small wooden blocks, in such a position that the pointer swings very close to it but does not touch it.

The meter is now complete except for providing it with binding posts connected to the terminals of the coil.

Before the meter can be used it must be *calibrated* or have its scale properly marked. The method of doing this will be described farther on.

VOLTMETER AND AMMETER

The meter illustrated in Fig. 112 is enclosed in a case. Otherwise it is quite similar to the one just described.

The bobbin upon which the wire is wound is made of cigar box wood. In laying out the work, scratch the lines on the wood with the point of a sharp penknife. Pencil lines are too thick to permit accuracy in small work. When finished, the bobbin must be perfectly true and square. The dimensions may be secured from the illustration. In putting the bobbin together, do not use any nails. Use strong glue only. Two bobbins are required, one for the ammeter and one for the voltmeter.

The bobbin for the ammeter is wound with No. 14 B. & S. enameled magnet wire wound on in smooth even layers. The voltmeter requires No. 40 B. & S. enameled wire. Such fine wire is difficult to wind smoothly but do the best you can.

A small hole is bored in the flange through which to pass the end of the wire when starting to wind. After finishing the winding about six inches of wire should be left at both ends to make connection with the terminals.

The magnetic needle or *armature* is a piece of soft steel one inch long, one-eighth of an inch thick and three-eighths wide. A one-eighth inch hole is bored one-sixteenth of an inch above the center to receive the shaft. The center of gravity is thus thrown below the center of mass in the armature and the pointer will always return to zero if the instrument is level.

The shaft is a piece of one-eighth inch, Bessemer steel rod, seven-sixteenths of an inch long. The ends are filed to a sharp knife-edge on the under side as shown in the illustration.

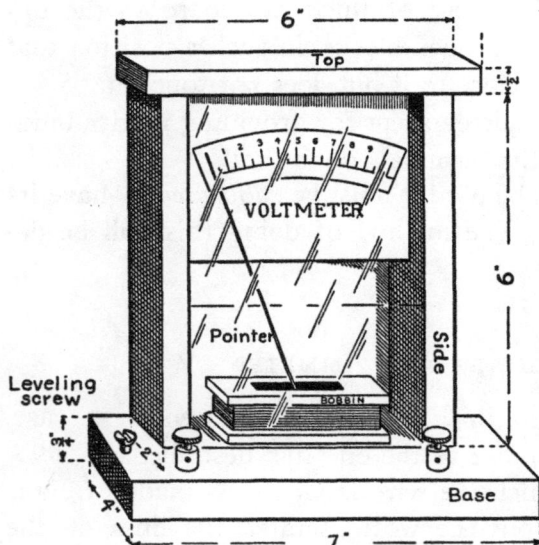

Fig. 112. — Completed Voltmeter.

A one-sixteenth-inch hole is bored in the top of the armature to receive the lower end of the pointer, which may be a piece of broom-straw, but is preferably a piece of No. 18 aluminum wire, four and one-half inches long.

After the holes have been drilled in it, the armature must be hardened so that it will retain its magnetism. It is heated to a bright red heat and dropped in a strong salt water solution. It is then magnetized by rubbing one end against the pole of a strong magnet or by winding ten or twelve turns of magnet wire around one end and connecting it with a battery for a moment.

The bearings are formed by two strips of thin sheet-brass, three-sixteenths of an inch wide, and one and one-quarter inches long, bent and cemented to the sides of the bobbin. In the illustration, part of the bobbin is shown as cut away so as to reveal the arrangement of the bearing. The center of the bearing is bent out so that the end of the shaft will not come in

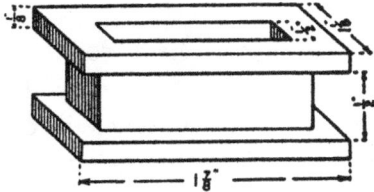

FIG. 113. — Details of the Bobbin.

contact with the sides of the bobbin. The top of the center is notched with a file to form a socket for the knife-edges on the shaft.

The bobbin is glued to the center of a wooden base, seven inches long, four inches wide and three-quarters of an inch thick. The terminals of the coil lead down through two small

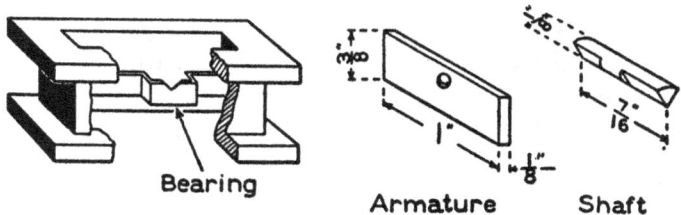

Bearing   Armature   Shaft

FIG. 114. — The Bobbin Partly Cut Away so as to Show the Bearing. Details of the Armature and Shaft.

holes in the base and thence to two large binding-posts. The wires are inlaid on the underside of the base, i.e., they pass from the holes to the binding posts through two grooves. This precaution avoids the possibility of a short-circuit or broken wire.

The case is formed of two sides, a back and a top all made of wood one-half an inch thick. It is six inches high, four inches wide, and two inches deep. A glass front slides in two shallow grooves cut in the sides, one-eighth of an inch from the front edge.

The case is held down to the base by four round-headed brass screws, which pass through the base into the sides. It is then easily removable in case it ever becomes necessary to repair or adjust the instrument. The top of the case is also arranged to lift off so that the armature may be replaced on its bearings should it happen to be thrown out of place by rough handling.

A brass screw, long enough to pass all the way through the base, serves to level the instrument. If a small brass strip is placed in the slot in the screw-head and soldered so as to form what is known as a "winged screw," the adjustment may be easily made with the fingers and will not require a screw-driver.

If the instrument is intended for mounting on a switch-board, it can be given a much better appearance by fitting with a smaller base, similar in size and shape to the top. The binding posts are then mounted on the sides.

To Calibrate the Meters

To calibrate the meters properly they must be compared with a standard meter. The scale is a piece of white cardboard glued to two small blocks on the inside of the case. The various values are marked with a pencil and then made permanent with a pen and ink.

The zero value on the scale will normally be in the center. When a current is passed through the bobbin, the armature tends to swing around at right angles to the turns of wire.

But since the armature is pivoted above its center of gravity, when it swings, its weight will exert a pull in opposition to the magnetic field of the bobbin. The deflection or swing of the pointer will be greater as the current is stronger. The pointer will swing either to the right or the left, depending upon the direction in which the current passes through the bobbin. The pointer of the instrument shown in the illustration is at the extreme left end of the scale. This is accomplished by bending

Voltmeter   Ammeter

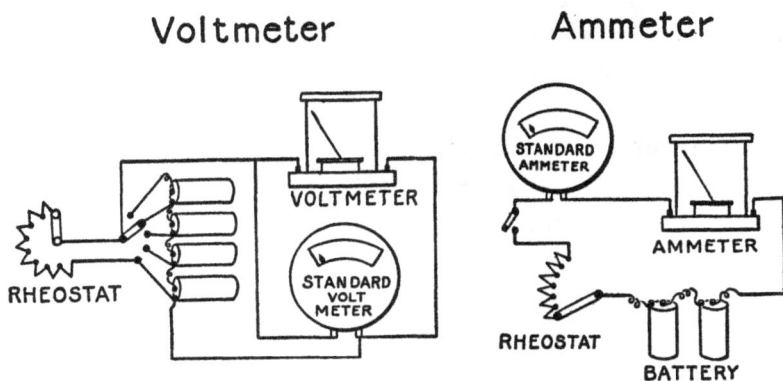

FIG. 115. — Circuits for Calibrating the Ammeter and Voltmeter.

the pointer to the left. The current value will be indicated when passing through the meter in only one direction, but the scale will have a greater range. In this case it will be necessary to cut a small groove in the base so that the armature will have plenty of room in which to swing.

When calibrating the *ammeter* it is placed in *series* with the standard meter, a set of strong batteries (a storage cell is preferred), and a rheostat. The rheostat is adjusted so that various current readings are obtained. The corresponding positions of the pointer on the meter being calibrated are then marked.

When calibrating a *voltmeter*, it must be placed in *parallel* or shunt with another voltmeter and in series with a battery. A switch is arranged so that the voltage of a varying number of cells may be passed through the meters. To secure fractional values of a volt, a rheostat is placed in shunt with the first cell of the battery. Then, by adjusting both the switch and the rheostat, any voltage within the maximum range of the battery may be secured.

This means of regulating or adjusting voltage by means of a rheostat connected across the terminals of a battery is common practice in laboratories. When so used, the rheostat is called a potentiometer.

When using the meters, it is always necessary that the ammeter shall be in series and the voltmeter in parallel or shunt with the circuit.

CHAPTER IX

BELLS, BURGLAR ALARMS AND ANNUNCIATORS

An electric bell may be purchased for twenty-five cents and from the standpoint of economy it does not pay to make one. But a bell is not a hard thing to construct, and the time

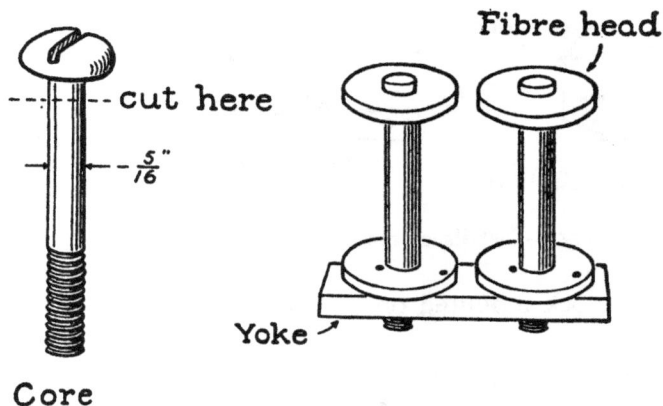

Fig. 116. — Details of the Magnet Spools and Yoke for an Electric Bell.

spent will be amply repaid by the more intimate knowledge of this interesting and useful piece of apparatus which will be gained by constructing it.

The wooden base is four inches wide and five and one-half inches long.

The electromagnets consist of two machine bolts, wound with No. 22 B. & S. gauge enamel or cotton-covered magnet wire. Fiber ends are fitted on the ends to hold the wire in place. Cover the cores with two or three layers of paper before winding on the wire. The terminals of the wire are led through holes in the fiber ends. The ends of the bolts are passed through the yoke, and the nuts applied to hold them in place. The magnets are then clamped down to the base by means of a hard-wood strip having a screw passed through it between the magnets and into the base.

The armature of the bell is made of a piece of iron having a flat spring riveted to it, as illustrated. The armature is fastened

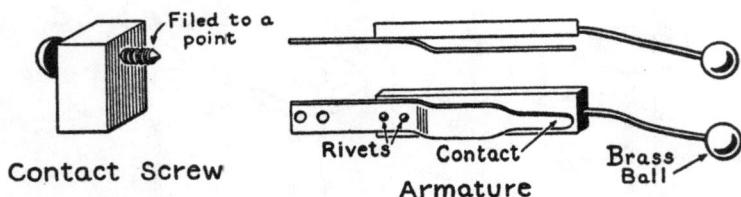

Fig. 117.—Details of the Armature and Contact Screw.

to a small block mounted on the lower left-hand corner of the base.

A second block with a contact-point made from an ordinary brass screw by filing the end to a point, is mounted on the base so that it is opposite the end of the contact spring fastened to the armature.

The gong may be secured from an old bell, alarm clock, or toy. It is mounted on the upper part of the base in such a position that the hammer will strike it on the lower edge.

The electromagnets are connected in series so as to present opposite poles to the armature. One terminal of the electromagnets is connected to the contact screw. The other is connected to one of the binding posts. A second binding post is connected to the armature.

The armature spring should be bent so that the armature is pushed over against the contact.

If a battery is connected to the bell, the electromagnets will pull the armature and cause the hammer to strike the gong.

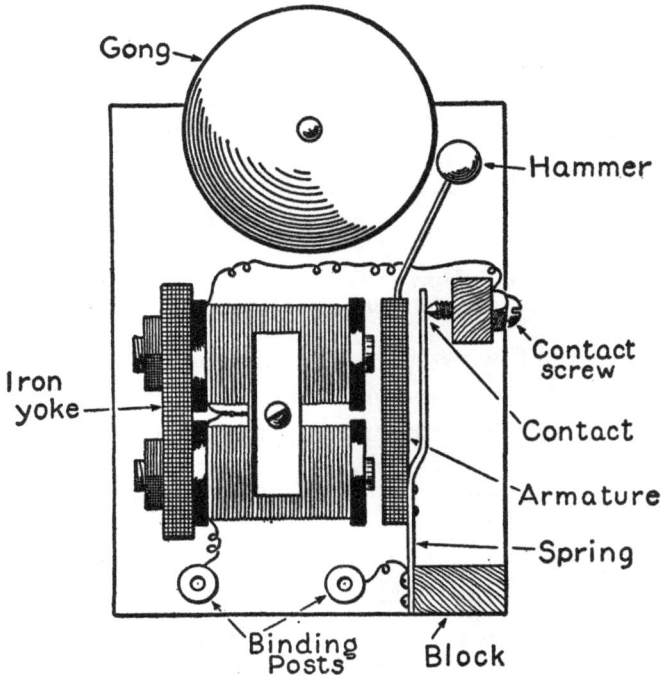

FIG. 118. — The Completed Bell.

As soon as the armature has moved a short distance, the spring will move away from the contact and break the circuit. The electromagnets cease pulling as soon as the current is cut off and the armature spring then causes the armature to move back and touch the contact. As soon as the contact is re-established, the armature is drawn in again and the process is repeated.

A little experimenting with the bell will soon reveal the best adjustment. Fig. 119 shows how to connect a bell, battery and push-button. A push-button is really a small switch which

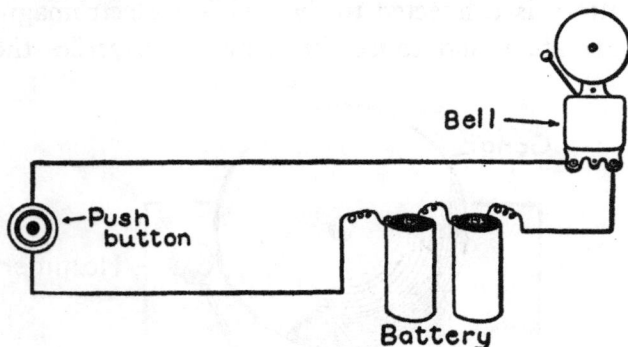

FIG. 119.—How to Connect a Bell, Battery and Push-Button.

closes the circuit when pressed. Do not make the armature spring too weak, or the hammer will move very slowly and with little life. When the armature moves toward the magnets, it

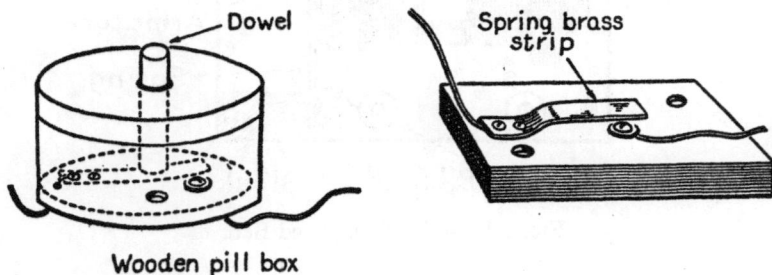

FIG. 120.—Two Simple Push-Buttons.

should barely touch the iron cores before the hammer strikes the gong.

Two simple methods of making push-buttons are illustrated.

It is sometimes desirable to arrange two bells and two push-buttons so that a return signal can be sent. In that case a circuit can be arranged so that a person can answer a bell and indicate that he has heard the call by pushing the second button. For example, one push-button and bell might be located on the top floor of a house and the other bell and button in the basement. A person in the basement wishing to call another on the

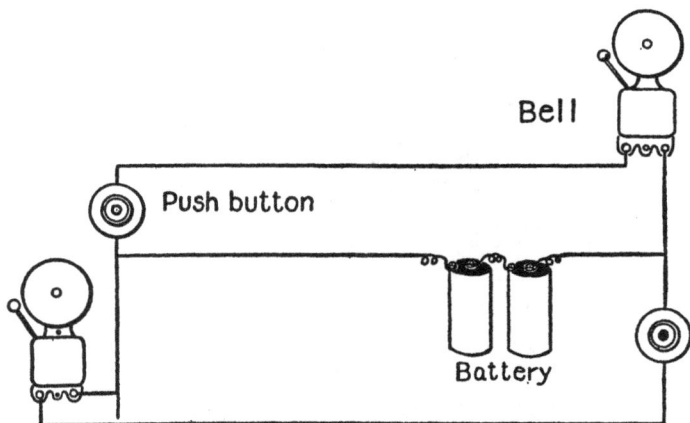

FIG. 121. — How to Arrange a Bell System of Return Signals.

top floor would push the button. The person answering could return the signal by pushing the button on the top floor and cause the bell in the basement to ring.

BELL-RINGING TRANSFORMERS

Bell-ringers, as they are sometimes called in an electrician's parlance, are small transformers which reduce the potential of the 110-volt lighting circuit to 6–8 volts. They will replace several dry cells as a source of current in a bell circuit.

A BURGLAR ALARM

A simple method of making an efficient burglar alarm is shown in Fig. 122. The base is a piece of wood about five by six inches and half an inch thick. A small brass strip, A, is fastened to the base by means of two round-headed wood screws and the ends turned up at right angles. The lever, B, is also a strip of brass. One end is bent up, so as to clear the strip and the screws that are under it. The lever is pivoted in the center

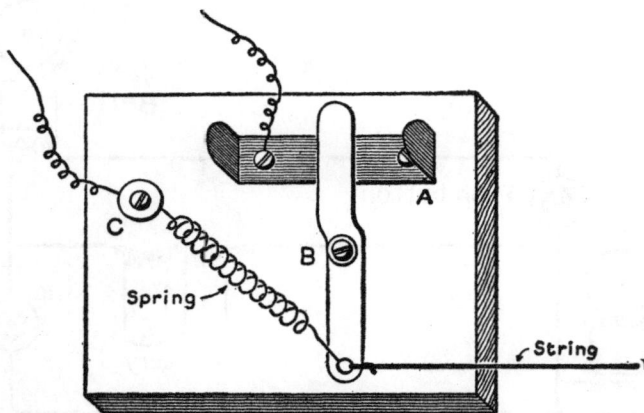

FIG. 122. — Burglar-Alarm Trap.

with a screw and washer. A small hole, D, is bored in the lower end so that a spiral spring may be attached. The other end of the spring is fastened under the screw and washer indicated by C in the illustration.

In order to set the alarm, fasten it down. Attach a string to the lever and stretch it across a window or doorway. Fasten the string in such a position that the lever is held half-way between the two ends of the strip, A.

If the string is disturbed, it will pull the lever over against the strip, A. If the string is cut, the spring will pull the lever over to the opposite side. In either case, if the alarm is connected to a bell and a battery, the circuit will be closed and the bell will ring.

One wire leading from the bell and the battery should be connected to A, and the other to the screw and washer C.

A black thread may be substituted for the string. If it is

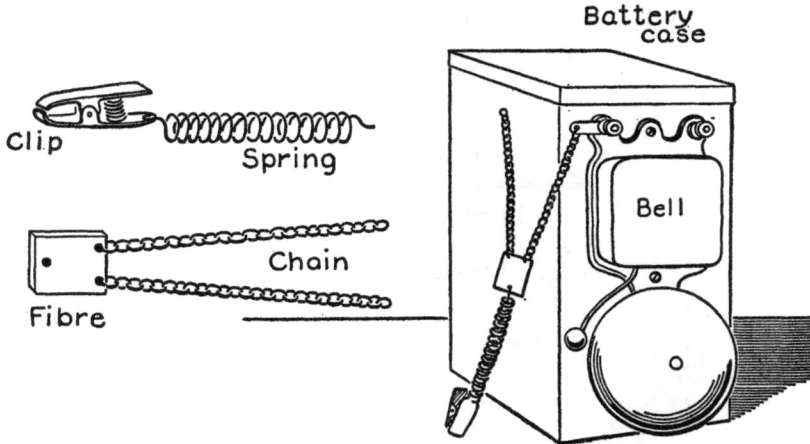

FIG. 123. — Details of the Chain Electrodes, etc. An Electric Alarm Attachment for a Clock.

stretched across a window or doorway an intruder entering in the dark and unaware of the existence of the alarm, would probably break the thread and ring the bell.

An Electric Alarm

It is often desirable to arrange an electrical alarm clock so that a bell will ring continuously until shut off.

Fig. 123 shows an electrical alarm attachment. It consists of a wooden box, large enough to contain an ordinary dry cell. An electric door-bell is fastened on the outside of the box. Connect one terminal of the battery to one terminal of the bell. Connect the other bell and battery terminals, each to a short piece of brass chain about four inches long. The ends of the chain are then fastened to a small piece of sheet fiber or hard rubber, so that they are insulated from each other. The opposite end of the fiber is fastened to a piece of wire spring having a clip soldered to the end.

The operation of this arrangement is very simple. Wind up the alarm key of an ordinary alarm clock and place the clip on

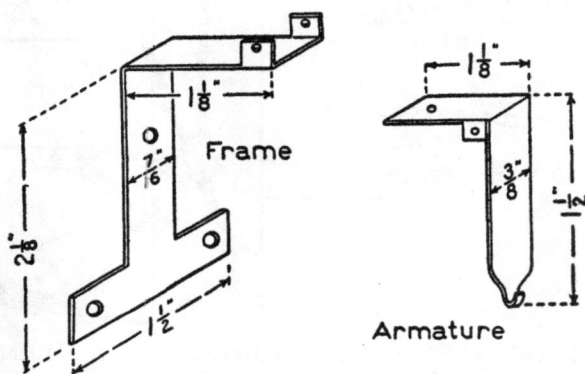

FIG. 124. — Details of the Drop-Frame and Armature.

the key. Place the clock in such a position that the two chains do not touch each other. Set the clock. When the mechanical alarm goes off, the key will revolve and twist the two chains together, thus closing the electric circuit and causing the bell to ring. The bell will ring until the clamp is removed. The alarm can be attached to any ordinary alarm clock.

AN ANNUNCIATOR

Annunciators are often placed in bell and burglar alarm circuits to indicate where the button ringing the bell was pushed, in case there are several.

The separate indicators used on an annunciator are called

drops. A drop may be made from an electromagnet and some sheet-metal.

The frame is cut from sheet-brass. The drop bar is also sheet-brass. It is pivoted on the frame at its lower end and has the upper end turned up to receive a numeral or letter.

The armature is made from a strip of sheet-iron. It is pivoted on the frame at its upper end. The strip is bent at right angles so as to fall in front of the magnet. The lower part of the armature is bent into a hook. The hook fits into a slot cut in the drop bar. A fine wire spring is placed between the frame and the upper end of the

FIG. 125. — An Annunciator Drop.

armature so as to pull the armature away from the magnet when the current is not flowing.

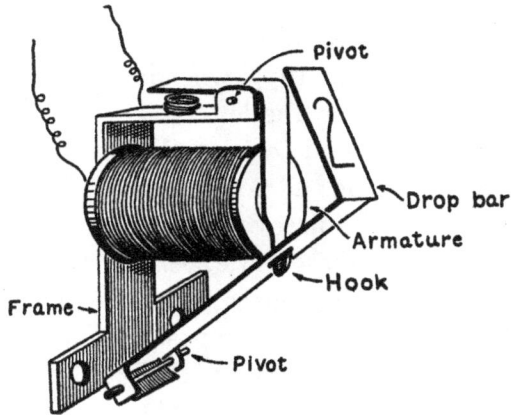

The electromagnet should be wound with No. 25 B. & S. gauge magnet wire.

When a current is sent through the electromagnet, it will draw the armature in. This action releases the drop bar and permits it to drop and bring the letter or number into view. A number of drops may be arranged on a board and placed in different circuits so as to indicate which circuit is closed at any time. It is a good plan to arrange a bar to act as a stop, so that the numerals will not fall past a horizontal position. Each time that one of the drops falls, it must be reset by pushing the bar back into position.

Chapter X

TELEGRAPHY

EXPERIMENTS in electric telegraphy were carried on as far back as the year 1753, when it was proposed to transmit messages by representing the letters of the alphabet by combinations of sparks produced by a static machine; but these were of little practical value and nothing of any importance was accomplished until after the discovery of the Voltaic cell.

Many of these old experiments were very crude and appear somewhat ridiculous when compared with the modern telegraph. The earliest known proposal for an electric telegraph appeared in the *Scots' Magazine* for February, 1753, and shows several kinds of proposed telegraphs acting by the attractive power of static electricity. It was proposed to convey static electricity from the sending station to the receiver by a series of parallel wires, one wire corresponding to each letter of the alphabet and supported by glass rods at every twenty yards. Words were to be spelled by the action of the electricity in attracting paper letters, or by striking bells corresponding to letters.

The modern telegraph system consists of a transmitting instrument much like a typewriter in appearance and a teletype receiver. Messages are sent by tapping keys corresponding to the different letters of the alphabet and received printed in Roman characters on a sheet of paper or tape.

The Morse system, which the teletype has almost completely replaced, consists essentially of four things, namely:

A battery which produces an electric current.

A wire which conducts the electric current from one point to another.

A transmitter for shutting the current off and on.

An electromagnetic receiving apparatus which gives out in

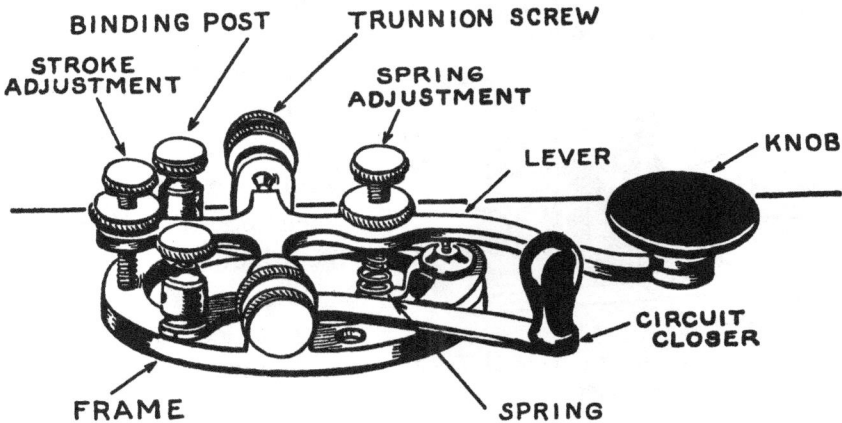

BINDING POST TRUNNION SCREW
STROKE ADJUSTMENT SPRING ADJUSTMENT
LEVER KNOB
CIRCUIT CLOSER
FRAME SPRING

FIG. 126. — Parts of a Typical Telegraph Key.

sounds the signals made by the pulsations of the current from a distant point.

The operation of the Morse telegraph is not complicated or difficult.

The Morse key is a contrivance for switching a current on and off. It consists of a steel lever, swung on trunnion screws mounted in a frame, and provided with a rubber knob which the operator grasps lightly with the thumb and forefinger. On pressing the lever downward, a platinum point fastened to the underside of the lever is brought into contact with another point set into a rubber bushing in the base of the key.

There is no electrical connection between the two points unless the key is pressed down or "closed" as it is often termed.

The lever is fitted with screws which permit the stroke or motion of the key to be closely adjusted.

The "sounder," line, wire and battery are connected to the key so that no current can flow until the key is pressed and the contacts brought together.

A sounder consists of two electromagnets mounted on a base under a movable flat piece of iron which is attracted by the magnetism of the electromagnets when a current flows through them and is withdrawn by a spring when no current excites the windings.

The movable piece of iron, called the armature, is mounted upon a strip of brass or aluminum called the *lever*. The lever strikes against a brass *anvil* and produces the "clicks" which form the dots and dashes of the telegraph alphabet.

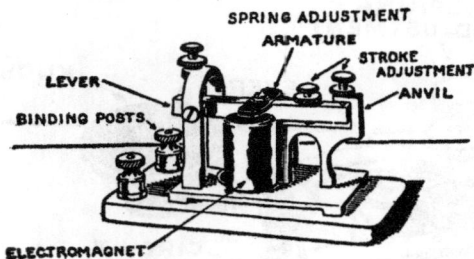

FIG. 127. — Parts of a Typical Telegraph Sounder.

Every time the key is pressed, an electric current is sent through the line. The current passes through the magnets of the sounder and causes the armature to be drawn downward. The lever strikes the anvil and produces a "click." When the key lever is released, the current is shut off and the lever flies up and clicks against the top of the anvil.

The period of time between the first click and the second click may be varied at will according to the length of time that the key is held down. A short period is called a *dot* and a long period a *dash*. Combinations of dots, dashes and spaces arranged according to the Morse alphabet make intelligible signals.

How to Make a Simple Key and Sounder

The little telegraph instruments shown in Figs. 128 and 129 are not practical for long lines but may be used for ticking messages from one room to another, and can be made the source of much instruction and pleasure.

The key is a strip of spring brass fastened to a wooden base in the manner shown in Fig. 128. It is fitted with a flat knob

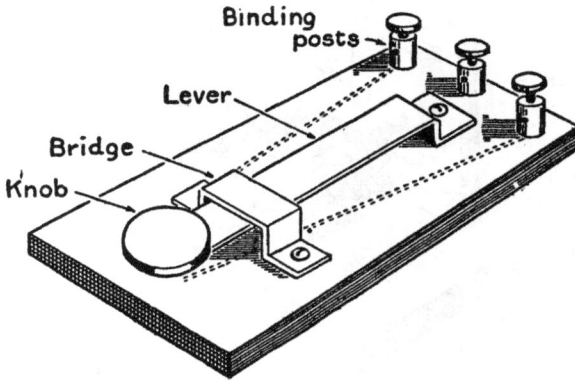

Fig. 128. — A Simple Home-Made Telegraph Key.

of some sort on the front end, so that it is conveniently gripped with the fingers.

The little bridge is made from sheet-brass and prevents the lever from moving too far away from the contact on the upward stroke.

Connections are made to the key lever at the back end and the contact in front by binding-posts. The center binding-post connects with the key lever. One of the outside binding-posts is connected to the bridge and the other to the contact.

The sounder consists of two small electromagnets mounted in a vertical position on a wooden base. The magnets are connected at the bottom by a strip of sheet-iron which acts as a yoke.

The armature is made out of sheet-iron, folded over in the manner shown in the illustration. One end of the armature is

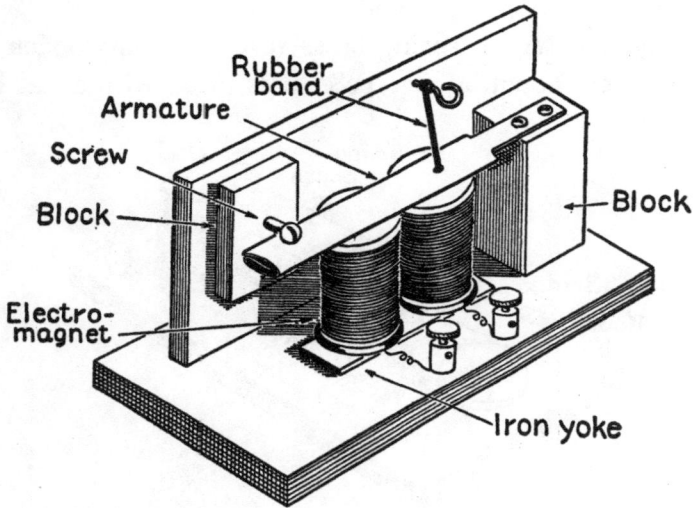

Fig. 129. — A Simple Home-Made Telegraph Sender.

fastened to a wooden block in such a position that the armature comes directly over the magnets and about one-eighth of an inch above them. The opposite end of the armature moves up and down for about an eighth of an inch between two screws, each fastened in a wooden block mounted on an upright board in the back of the magnets. The purpose of the screws is to make the "click" of the sounder louder and clearer than it would be if the armature only struck the wood.

A rubber band or a small wire spring passing over a screw-eye and connected at the other end to the armature will draw the latter away from the magnets when the current is not passing.

The terminals of the magnets are connected to binding-posts mounted on the base.

The key and sounder should be placed in series with one or two cells of a battery. Pressing the key will then cause the armature of the sounder to be drawn down and make a click. When the key is released, the armature will be drawn up by the spring or rubber band and make a second click.

What boy interested in mechanics and electricity has not at

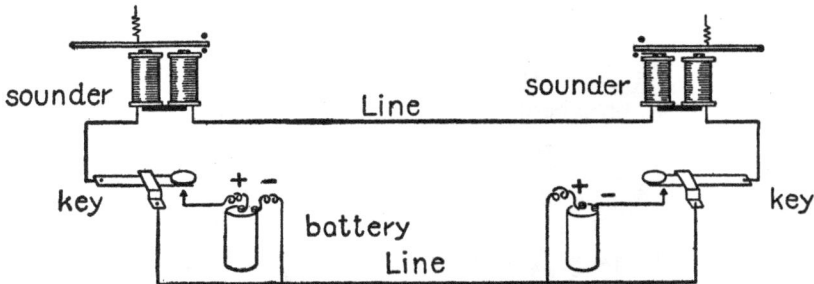

FIG. 130. — How to Connect Two Simple Telegraph Stations.

some time or other wished for a telegraph instrument with which to put up a "line" with his chum?

A practical set of such instruments can be easily constructed, and with little expense.

The electromagnets for the sounder may be constructed by the intending telegraph operator or secured from some old electrical instrument such as a magneto-bell. In the latter case, the hardest part of the work will be avoided.

If they are to be home-made, the following suggestions may prove of value.

The cores are made from five-sixteenths-inch stove-bolts with the heads cut off. The magnet heads are cut out of fiber, one-sixteenth of an inch thick and one inch in diameter. They should fit tightly and be held in place with celluloid cement. They are separated so as to form a winding space of seven-eighths of an inch. The magnets should be wound full of No. 25 B. & S. gauge cotton-covered wire.

The yoke may be solid or made of enough strips of sheet-iron, three-quarters of an inch wide and two inches long, to form a pile one-quarter of an inch thick. Two five-sixteenths-inch holes are bored in the opposite ends of the yoke, one and

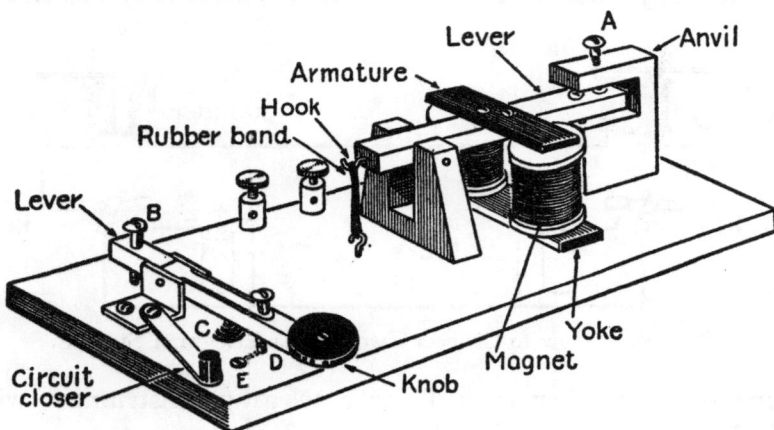

FIG. 131. — A Complete Telegraph Set, Consisting of a Keyboard and a Sounder.

one-half inches apart. The lower ends of the magnet cores are passed through these holes. The ends should project one-half inch beyond the yoke.

They are passed through two holes in a base-board three-quarters of an inch thick. The holes are counter-sunk from the lower side, so that a nut can be screwed on the lower end of each and the magnets held tightly in an upright position. The remaining parts of the instrument are easily made, and are so clearly shown by the drawing that it is hardly necessary to say more than a few words in explanation.

The lever, the anvil, the standard and the lever of the key are all cut out of hard-wood according to the pattern shown in the illustration.

The armature is a piece of soft iron fastened to the lever with a small brass screw.

Tacks are placed under the heads of the adjusting screws on the sounder so that it will click more loudly.

The rubber band acts as a spring to counteract the weight of

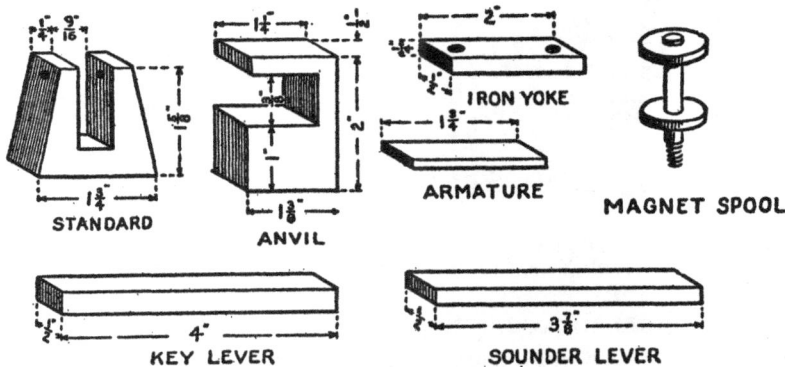

STANDARD

ANVIL

IRON YOKE

ARMATURE

MAGNET SPOOL

KEY LEVER

SOUNDER LEVER

Fig. 132. — Details of the Telegraph Set Shown in Figure 131.

the armature and lever and draw it up as soon as the current is cut off. The movement of the lever should be so adjusted that it is only sufficient to make an audible click.

Use care to avoid friction between the lever and the standard, so that the former will move with perfect freedom.

All the screws used in the work should be round-headed brass wood screws with the points filed flat. Bore a small hole before screwing them into place so as to avoid splitting the wood.

The construction of the key is even more simple than that of the sounder. It should move up and down without any side motion.

The circuit-closer should be kept closed when the instruments are not in use, and when you are receiving a message. As soon as you are through receiving and wish to transmit, you should open your circuit-closer and your friend close his.

The tension of the spring under the lever of the key must be adjusted to suit the hand of each individual operator.

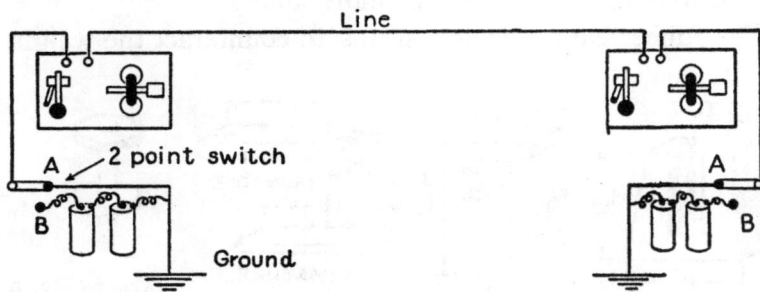

FIG. 133. — A Diagram Showing How to Connect Two Complete Telegraph Sets, Using One Line Wire and a Ground.

The diagram for connecting the instruments is self-explanatory. In cities or towns where a "ground" is available by connecting to the gas or water pipes, one line wire may be easily dispensed with. Or, if desirable, a ground may be formed by burying a large plate of zinc (three or four feet square) in a moist spot and leading the wire to it.

How to Build a Telegraph Relay

In working a telegraph over a long line or where there are a large number of instruments on one circuit, the currents are often not strong enough to work the sounder directly. In such a case a *relay* is used. The relay is built on the same principle as a sounder, but the parts are made much lighter, so that the instrument is more sensitive. The armature of a relay is so light and its movement so small that its clicking is scarcely audible. It is therefore fitted with a second set of contacts and connected to a battery and a sounder, which is set in operation every time the contacts close. The principle of a relay is that a weak

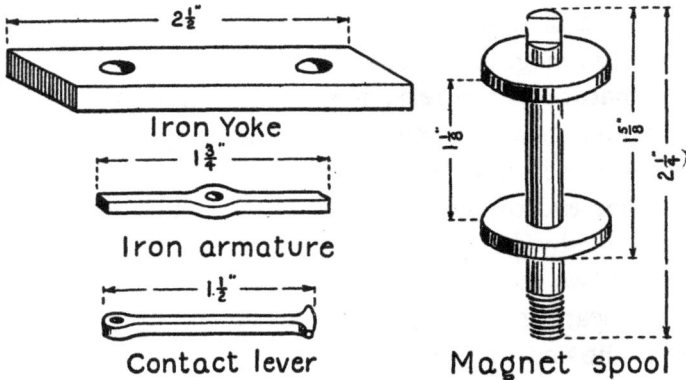

Fig. 134. — Details of the Relay Parts.

current of insufficient strength to do the work itself may set a strong local current to do its work for it.

There are many forms of relays, and while that which is described below is not exactly the type commonly used on telegraph lines, it has the advantage of being far more sensitive than any instrument of the regular line relay type that the average experimenter could build.

Make the magnets from one-quarter-inch stove-bolts, and cut them off so that they will form a core about two and one-quarter inches long. Fit each of the cores with two fiber heads to hold the wire in place. Insulate the cores with paper and wind each with about fifty layers of No. 30 to 36 B. & S. gauge single-cotton-covered magnet wire. The winding space between the magnets' heads should be one and one-eighth inches.

The upper ends of the magnet cores should be allowed to project about one-quarter of an inch beyond the fiber head. The end of the core is filed flat, as shown in the illustration.

The magnets are mounted upon an iron yoke, three-sixteenths of an inch thick. The holes in the yoke should be spaced so that there is a distance of one and one-half inches between the centers of the magnet cores.

The armature of the relay is mounted on a small steel shaft with sharp points at each end. The exact shape of the armature may be best understood from the illustrations.

The lower end of the shaft rests in a small cone-shaped depression made by driving a center punch into the yoke half-way between the two magnets.

The top bearing is a strip of brass projecting from a wooden support. The end of the shaft rests in a depression similar to that in the yoke.

The contact lever is made of brass and forced on the shaft below the armature. It swings between a small brass clip fastened to one side of the support and a little screw held in a similar piece on the opposite side.

The contact is made of brass about No. 22 gauge in thickness. It may be adjusted by a screw passing through the support.

The armature may be controlled in its movement so that the latter will be very slight by adjusting the screws.

There should not be any friction in the bearings and the armature should move with perfect freedom. The armature should approach the ends of the magnet cores until a space about the thickness of heavy paper separates them and should not touch them.

The spring is made of fine brass wire. It is fastened to the armature shaft, and the screw mounted on the wooden support with a piece of silk thread. The thread is passed around the shaft once or twice so that the tension of the spring will cause the armature to move away from the pole pieces just as soon as the current flowing through the magnets ceases.

The tension of the spring may be adjusted by turning the

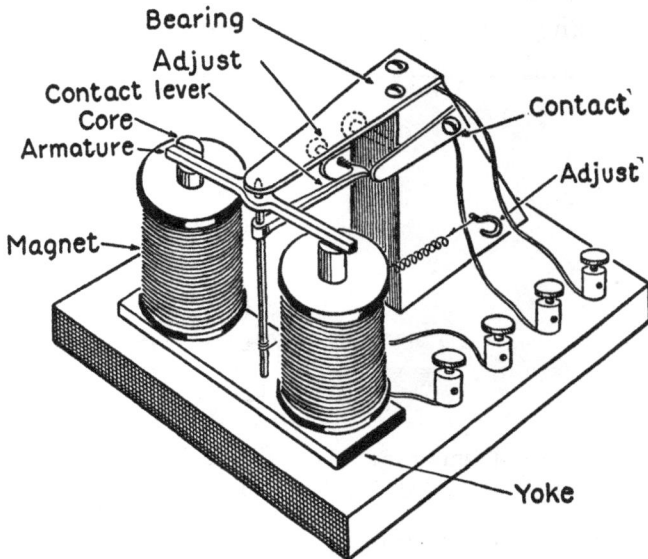

FIG. 135. — The Completed Relay.

screw-eye. If the armature tends to stick to the magnet poles, fasten a small piece of paper to the poles with some shellac.

The terminals of the magnets are connected to two binding-posts marked A and B. The binding-posts marked C and D are connected respectively to the contact clip and the brass bearing on the top of the wooden support.

The diagram in Fig. 136 shows how the relay is connected to a telegraph line.

How to Learn to Telegraph

The instruments so far described are practical telegraph instruments, but they lack the fine points of commercial apparatus and it is not possible to become as efficient an operator with their aid as with a real key and sounder.

If the young experimenter desires to become a proficient

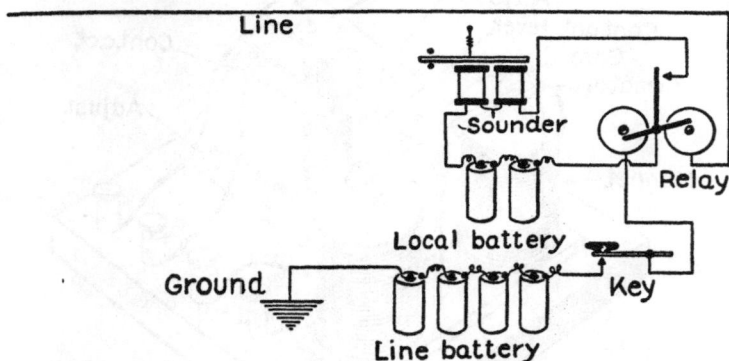

FIG. 136. — A Diagram Showing How to Connect a Relay, Sounder, and Key.

telegraph operator, the first thing to do is to purchase a Learner's telegraph key and sounder.

Connect a dry cell to the binding-posts on the back of the instrument. Screw the set down on a table about eighteen inches from the front edge, so that there is plenty of room for the arm to rest. See that none of the adjustment screws about the instrument are loose and that the armature of the sounder moves freely up and down through a distance of about one-sixteenth of an inch.

The spring which draws the lever upwards away from the magnets should be set only at sufficient tension to raise the lever when no current is passing. If too tight, the spring will not allow the armature to respond to the current flowing through the magnets.

The key is provided with two adjustment-screws to regulate the tension and the play of the lever to suit the hand of the

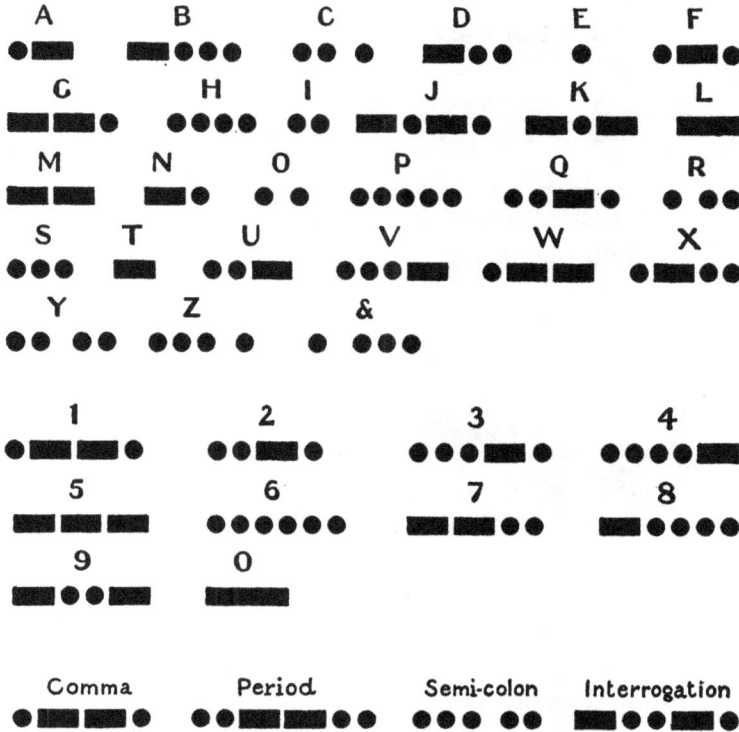

FIG. 137. — The Morse Telegraphic Code.

operator. A little practice will enable the student to judge best for himself just how the key should be set.

The next step is to memorize the alphabet, so that each character can instantly be called to mind at will. The punctuation marks are not used very frequently, and the period is the only one which the student need learn at first.

The Morse alphabet consists of dots, dashes, and spaces. Combinations of these signals spell letters and words.

Many of the characters are the reverse of others. For example, A is the reverse of N, B and V, D and U, C and R, Q and X, Z and &, are the other reverse letters, so if the formation of one of each of these letters is memorized the reverse is easily mastered.

It is important that the beginner should learn how to grasp the key properly, for habits are easily formed and a poor position will limit the sending speed of the operator.

Place the first or index finger on the top of the key-knob, with the thumb under the edge; and the second finger on the

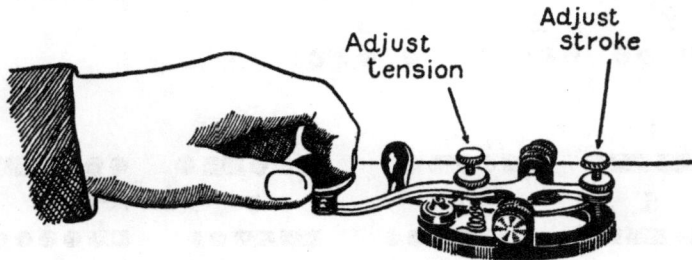

Fig. 138. — How to Hold a Telegraph Key.

opposite side. The fingers should be curved so as to form a quarter-section of a circle. Bring the third and fourth fingers down so that they are almost closed on the palm of the hand. Rest the arm on the table in front of the key and allow the wrist to be perfectly limber.

The grasp on the key should be firm but not rigid. Avoid using too much strength or a light hesitating touch. Endeavor to acquire a positive, firm up and down motion of the key. Avoid all side pressure, and do not allow the fingers to leave the key when making the signals. The movement is made principally with the wrist, with the fingers and hand perfectly elastic.

A dot is made by a single instantaneous, downward stroke of the key. A dash is made by holding the key down for the same period of time that it takes to make three dots. A long dash is made by holding the key down for the same time that it takes to make five dots.

A space in the letters, such as, for instance, the space between the first and last two dots in the letter R should occupy the time of one dot. The space between each letter should occupy the time required for two dots, and the space between words should occupy the time required for three dots.

Commence to use the key by making dots in succession, first at the rate of two every second, and increasing the speed until ten can be made. Practice should be continued until three hundred and sixty dots a minute can be made with perfect regularity.

Then begin making dashes at the rate of two every three seconds, and continue until one hundred and twenty a minute can be made with perfect regularity.

Practise the long dashes at the rate of one a second, and increase until ninety can be made in a minute.

When this has been accomplished, practise the following letters until they can be perfectly made. Each row of letters is an exercise which should be practised separately until mastered.

DOT LETTERS

E   I   S   H   P   6

DOT AND SPACE LETTERS

O   C   R   Y   Z   &

DASH LETTERS

T   L   M   5   0

DOTS AND DASHES

A   U   V   4

DASHES AND DOTS
N D B 8
MIXED DOTS AND DASHES
F G J K Q W
X 1 2 3 7 9 Period

After you can write these different letters, practise making words. Select a list of commonly used words. When words seem easy to write, practise sending pages from a book.

Systematic and continual practice will enable the student to make surprising progress in mastering the art of sending.

Reading and receiving messages must be practised with a companion student. Place two instruments in separate rooms or in separate houses so that the operators will be entirely dependent upon the instruments for their communication with each other. Start by transmitting and receiving simple messages. Then use pages from a book, and increase the speed until it is possible to send and receive at least 15 words a minute without watching the sounder but merely depending upon the clicks to determine the duration of the dots and dashes.

Fig. 133 shows how to arrange a regular telegraph line for two stations. Gravity batteries should be used for regular telegraph work. It is necessary that the key should be kept closed by having its circuit-closer shut when messages are not being sent. If one of the keys is left open the circuit is broken, and it is not possible for a person at the other end of the line to send a message.

Every telegraph office has a name or call usually consisting of two letters; thus for New York the call might be N. Y. and for Chicago, C. H.

When one station desires to call another the operator repeats the call letters of that station and signs his own call until answered.

CHAPTER XI

MICROPHONES AND TELEPHONES

IN 1878, David Edward Hughes discovered that the contact formed between two pieces of some such substance as carbon or charcoal is very sensitive to the slightest changes in pressure, and when included in an electric circuit with a battery and a telephone receiver, will transmit sounds. Such an instrument is called a *microphone*. It has various forms but in most of them one piece of carbon is held loosely between two other pieces in such a manner as to be easily affected by the slightest vibrations conveyed to it through the air or any other medium.

A simple form of instrument embodying this principle consists of a small pencil of carbon supported loosely between two blocks of the same substance glued to a thin wooden sounding-board. The sounding-board is mounted in an upright position on a wooden base. The carbon pencil rests loosely in two small indentations in the carbon blocks. The blocks are connected, by means of a very fine wire or a strip of tinfoil, with one or two cells of battery and a telephone receiver. Vibration or sounds in range of the microphone will cause the sounding-board to vibrate. This affects the pressure of the contact between the carbon pencil and the two blocks. When the pressure between the two is increased the resistance in the path of the electric current is decreased and more current immediately flows through the circuit. On the other

173

hand, when the pressure is decreased, the resistance is increased and less current flows through the telephone receiver. The amount of current flowing in the circuit thus keeps step with the changes in the resistance, and accordingly produces sounds in the telephone receiver. The vibrations emitted from the receiver are usually much greater than those of the original sounds, and so the microphone may be used to magnify weak sounds such as the ticking of clock-wheels or the footfalls of insects. If a watch is laid on the base of the microphone, the ticking of the escapement wheel can be heard with startling loudness. The sounds caused by a fly walking on a microphone may be made to sound "as loud as the tramp of a horse."

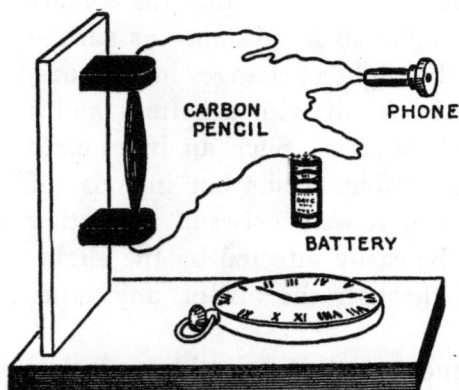

FIG. 139. — A Microphone Connected to a Telephone Receiver and a Battery.

The electrical *stethoscopes* used by physicians to listen to the action of the heart are in principle a microphone and telephone receiver connected to a battery.

It is easy to make a very sensitive microphone. With this instrument it is possible to hear the tramping of a fly's feet and the sounds of its wings.

The base upon which the apparatus is mounted serves as the sounding-board and is made in the form of a hollow wooden box. It can be made from an ordinary cigar-box by removing the paper and taking the box apart. The piece forming the top of the box must be planed down until it is only three thirty-seconds of an inch thick. The box should measure about six inches square and three-quarters of an inch thick when finished. Do not use any nails or small brads in its construction, but fasten it together with glue. If you use any nails you will decrease the sensitiveness of the instrument appreciably. The bottom of the box should be left open. The result is a sounding-board of the same principles as that of a banjo head. Small wooden feet, one-quarter of

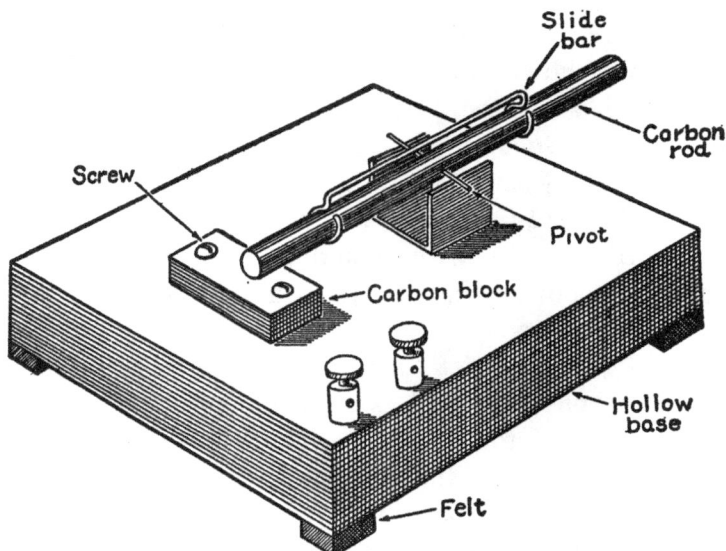

FIG. 140. — A Very Sensitive Form of Microphone, with which the Footsteps of a Fly Can Be Heard.

an inch square, are glued to the four under corners so as to raise the bottom clear of the table, or whatever the microphone may be placed upon. The bottom of each one of the small feet is cushioned with a layer of felt so that no jars will be transmitted to the instrument by any object upon which it is resting.

The carbon pencil used on this type of instrument is pivoted in the center and rests at one end upon a carbon block.

The carbon block is made about one inch long, one-quarter

of an inch thick, and one-half of an inch wide. A small hole is drilled near each end to receive a screw which fastens the block to the sounding-board. A fine wire is led from one of these screws to a binding-post mounted at the side of the box. Another wire leads from a second binding-post to a standard which is also fastened to the sounding-board with a small screw.

The standard is made from a sheet of thin brass and is bent into the shape shown in the illustration.

The pencil is a piece of one-quarter-inch carbon rod three inches long. A small hole is drilled through the pencil at a point one and five-eighths of an inch from one end with a sewing-needle, and a piece of fine brass wire, pointed at both ends, pushed in. The wire should be a tight fit in the hole. It should be about one-half of an inch long, and may be made from an ordinary pin.

The slide-bar is used to regulate the pressure of the pencil upon the carbon block and is simply a piece of soft copper wire. It is bent into the shape shown in the illustration so that it will slide over the carbon pencil. The sides of the standard should press just tightly enough against the ends of the pivot which passes through the carbon pencil to hold it in position without slipping, and at the same time allow it to swing freely up and down.

The two binding-posts should be connected in series with two dry cells and a pair of sensitive telephone receivers. Place the receivers against the ears. Move the slide-bar gently back and forth until the voice of any one talking in another part of the room can be heard in the telephone receivers. In order to hear faint whispers, move the slide-bar away from the carbon block.

In order to hear a fly walk it is necessary to have the carbons very dry and clean. The instrument must be very carefully adjusted. Cover the microphone with a large glass jar and place a fly inside the jar. Whenever the fly walks on any part of the microphone you will be able to hear each footstep in the telephone receivers. When he flies about inside of the globe, his wings will cause a roaring and buzzing noise to be heard in the receivers.

TELEPHONES

When the telephone made its first appearance, it was the wonder of the times just as radio telephony is to-day. Starting as an exceedingly simple and inexpensive apparatus, it has gradually developed into a wonderful and complex system, so that at the present time, instead of experiencing difficulty in telephoning over distances of fifty or one hundred miles, as at first, it is possible to carry on a conversation with almost any other part of the world.

Like the telegraph, the principle of the telephone is that of a current of electricity flowing over a line wire into a pair of electro-magnets, but with many important differences.

Compared with telegraph apparatus, a telephone is exceedingly sensitive. A telegraph relay requires perhaps about one-hundredth of an ampere to work it properly. A telegraph sounder will require about one-tenth of an ampere, but a telephone receiver will render speech audible with less than a millionth of an ampere, and therefore may almost be said to be a hundred thousand times more sensitive than a sounder.

Another difference between the telephone and the telegraph lies in the fact that the currents flowing over a telegraph line do not usually vary at a rate greater than twenty or thirty times a second, whereas telephone currents change their intensity hundreds of times a second.

The telephone is an instrument for the transmission of speech to a distance by means of electricity, wherein the speaker

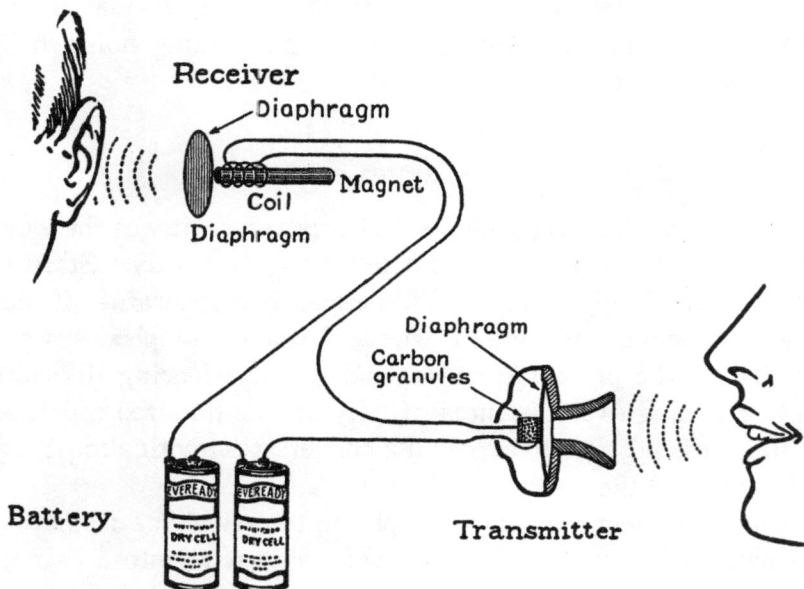

FIG. 141. — A Telephone System, Consisting of a Receiver, Transmitter, and a Battery Connected in Series.

talks to an elastic plate of thin sheet-iron which vibrates and sends out a pulsating current of electricity.

The transmission of the vibrations depends upon well-known principles of electricity, and does not consist of the actual transmission of sounds but of electrical impulses which keep perfect accord or step with the sound waves produced by the voice in the transmitter. These electrical currents pass through a pair of small electro-magnets acting upon a plate or diaphragm, which in turn agitates the air in a manner similar to the original voice speaking into the transmitter and thus emits sounds.

That part of the apparatus which takes up the sounds and changes them into electric currents composes the *transmitter.* When words are spoken into the mouthpiece they strike a diaphragm, on the back of which is fastened a small cup-shaped piece of carbon. A second cup is mounted in a rigid position directly back of the first. The space between them is filled with small polished granules of carbon. When these granules are in a perfectly loose state and are undisturbed, their resistance to an electric current is very great and they allow almost none to flow.*

When slightly compressed their resistance is greatly lowered and they permit the current to pass. The vibrations of the diaphragm cause the carbon cup mounted on its back to move and exert a varying pressure upon the granules with a corresponding variation in their resistance and the amount of current which will pass through.

The *receiver,* or that part of the apparatus which transforms the pulsating current back into sound waves, consists of a thin iron disk, placed very near but not quite touching the end of a small steel bar, permanently magnetized, and about which is wound a coil of fine insulated wire.

The transmitter and the receiver are connected together in series with a battery. When words are spoken into the transmitter the little carbon granules are immediately thrown into motion, and being alternately compressed and released cause corresponding changes in the current flowing through the receiver from the battery. The magnetism of the receiver changes with each change in the electric current, and thus by

---

* A transmitter is really a microphone built especially to receive the sounds of the human voice, and operates on the same principle.

alternately attracting and repelling the diaphragm causes it to
vibrate and emit sounds. Such is the *principle* of the telephone.
The telephones in actual service today are complicated with
bells, magnetos, induction coils, condensers, repeaters, relays
and various other apparatus, which renders them more efficient.

The bells and magnetos are for the purpose of calling the
central operator or the person at the other end of the line
and drawing attention to the fact that some one wishes to get
into communication with him. The older styles of telephones
used what is known as a polarized bell and a hand magneto for
this purpose. A polarized bell is a sensitive piece of apparatus
which will operate with little current. A magneto is a small
hand dynamo which when turned with a crank will generate
a current causing the bell at the other end of the line to ring.
When the telephone receiver is raised off its hook in order
to place it to the ear the bell and magneto are automatically
disconnected from the line and the receiver and the transmitter are connected in their place. The current necessary to supply the telephone and receiver is supplied by two or three
dry cells placed inside of each telephone.

The latest types of instruments employ what is known as
the central energy system, wherein the current is supplied by a
large storage battery located at the central office and serving
as a current supply to all the telephones connected to that system.

It would be impossible to explain the details of the telephone
systems in every-day use in such a book as this because of the
immense amount of material it would be necessary to present.
Such a work would occupy a volume of its own. Additional
information may be readily found in any reference library.
However, the "boy electrician" who wishes to make a telephone for communicating with his chum down the street,

will find the necessary instructions in the following pages. If this work is carried out carefully and a home-made telephone system installed it will not only prove a very interesting undertaking but will also serve to dispel all mystery which may surround this device in the mind of the young experimenter.

How to Build a Telephone

Telephone receivers are useful for many purposes in electrical work other than to receive speech. They are used in connection with radio instruments, in place of a galvanometer in measuring electrical circuits, and for testing in various ways.

Telephone receivers are of two types. One of them is long and cumbersome, and is very similar to the original Bell tele-

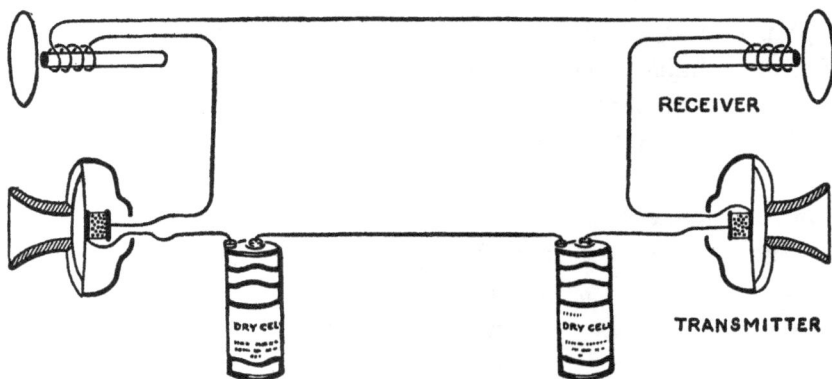

RECEIVER

DRY CELL

DRY CELL

TRANSMITTER

Fig. 142. — The Principle of a Home-Made Telephone.

phone receiver. The other is small and flat, and is called a "watch-case" receiver. A watch-case receiver consists of a U-shaped permanent magnet so mounted as to exert a polarizing influence upon a pair of little electro-magnets, before the poles of which is placed an iron diaphragm. For convenience, these parts are assembled in a small cylindrical casing, of hard-rubber or metal. The permanent magnet exerts a continual pull upon the diaphragm, tending to draw it in. When the telephone currents pass through the little magnets, they will either strengthen the permanent magnet and assist it in attracting the diaphragm, or detract from its strength and allow the diaphragm to recede, depending upon which direction the current flows.

Watch-case receivers are usually employed for radio work because they are very light in weight and can easily be attached to a head-band in order to hold them to the ears and leave the hands free. They are very useful to the amateur experimenter in many ways.

A telephone receiver capable of giving fair results on a short telephone line can be easily made, but of course will not prove as efficient as one which is purchased ready-made from a reliable electrical manufacturer.

The first practical telephone receiver was invented by Alexander Graham Bell and was made somewhat along the same lines as that shown here. If you have a wood-turning lathe you can make one.

Such a receiver may be made from a piece of curtain-pole, three and three-quarter inches long and about one and one-eighth inches in diameter. A hole, three-eighths of an inch in diameter, is bored along the axis throughout its entire length, to receive the permanent magnet.

The shell of the receiver is a cup-shaped piece of hard wood, two and one-half inches in diameter and one inch deep. It will have to be turned on your lathe. Its exact shape and dimensions are best understood from the dimensions shown in the illustration. The shell is firmly attached to one end of the piece of curtain-pole by gluing.

The permanent magnet is a piece of hardened steel, three-eighths of an inch in diameter and four and five-eighths of an inch in length. The steel will have to be tempered or hardened before it will make a suitable magnet. Accomplish this by heating the rod until red and then plunge it into water.

One end of the bar is fitted with two thick fiber washers about seven-eighths of an inch in diameter and spaced one-

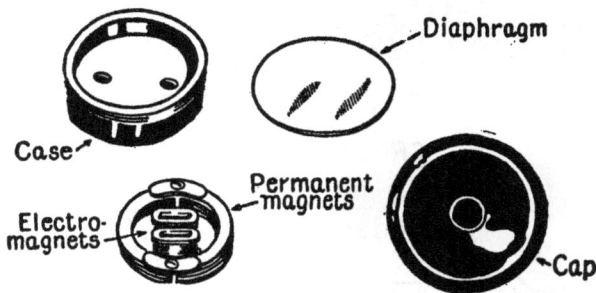

FIG. 143. — A Watch-Case Telephone Receiver.

quarter of an inch apart. The bobbin so formed is wound full of No. 36 B. & S. gauge single-silk-covered magnet wire. The ends of the wire are passed through two small holes in the fiber washers and then connected to a pair of heavier wires. The wires are run through two holes in the curtain-pole, passing lengthwise from end to end, parallel to the hole bored to receive the bar magnet.

This bar magnet is then pushed through the hole until the end of the rod on which the spool is fixed is just below the level of the edges of the shell.

The two wires are connected to binding-posts, A and B, mounted on the end of the receiver. A hook is also provided so that the receiver may be hung up.

The diaphragm is a circular piece of thin sheet-iron, two and one-half inches in diameter. It is placed over the shell, and the bar magnet adjusted until the end almost touches the diaphragm. The magnet should fit into the hole very tightly,

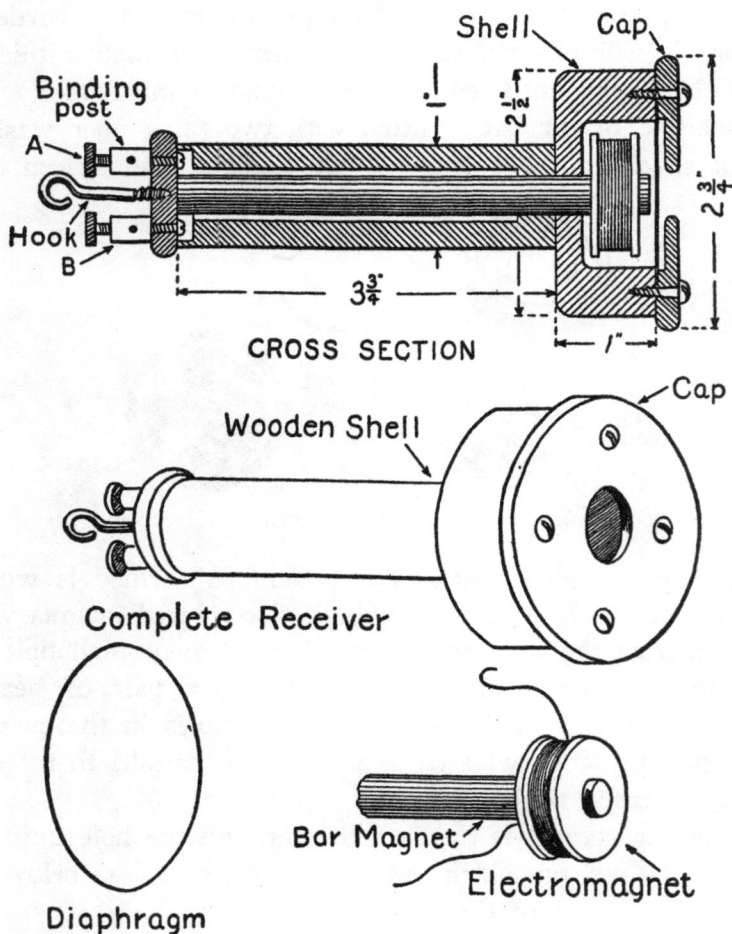

FIG. 144. — A Simple Form of Telephone Transmitter.

so that it will have to be driven in order to be moved back and forth.

The diaphragm is held in place by a hard-wood cap, two and

three-quarter inches in diameter and having a hole three-quarters of an inch in diameter in the center. The cap is held to the shell by means of four small brass screws.

The receiver is now completed and should give a loud click each time that a battery is connected or disconnected from the two posts, A and B.

The original Bell telephone apparatus was made up simply of two receivers without any battery or transmitter. In such a case the current is generated by "induction." The receiver is used to speak through as well as to hear through. This method of telephoning is unsatisfactory over any appreciable distances. The time utilized in making a transmitter will be well spent.

A simple form of transmitter is shown. The wooden back, B, is three and one-half inches square and three-quarters of an inch thick. The front face of the block is hollowed out in the center as shown in the cross-section view.

The face-plate, A, is two and one-half inches square and one-half an inch thick. A hole, seven-eighths of an inch in diameter, is bored through the center. One side is then hollowed out to a diameter of one and three-quarter inches, so as to give space for the diaphragm to vibrate as shown in the cross-sectional drawing.

The carbon buttons are about one inch in diameter and three-sixteenths of an inch thick. You can cut these from a carbon rod obtained from an old dry cell. A small hole is bored in the center of each to receive a brass machine screw. The hole is countersunk, so as to bring the head of the screw down as close to the surface of the carbon as is possible. Then, using a sharp knife or a three-cornered file, score the surface of the carbon until it is covered with criss-cross lines.

The diaphragm is a piece of thin sheet-iron cut in the form

of a circle two and one-half inches in diameter. A small hole is bored through the center of this. One of the carbon buttons is fastened to the center of the diaphragm with a small screw and a nut.

Cut out a strip of flannel or thin felt, nine-sixteenths of an inch wide and three and one-half inches long. Around the edge

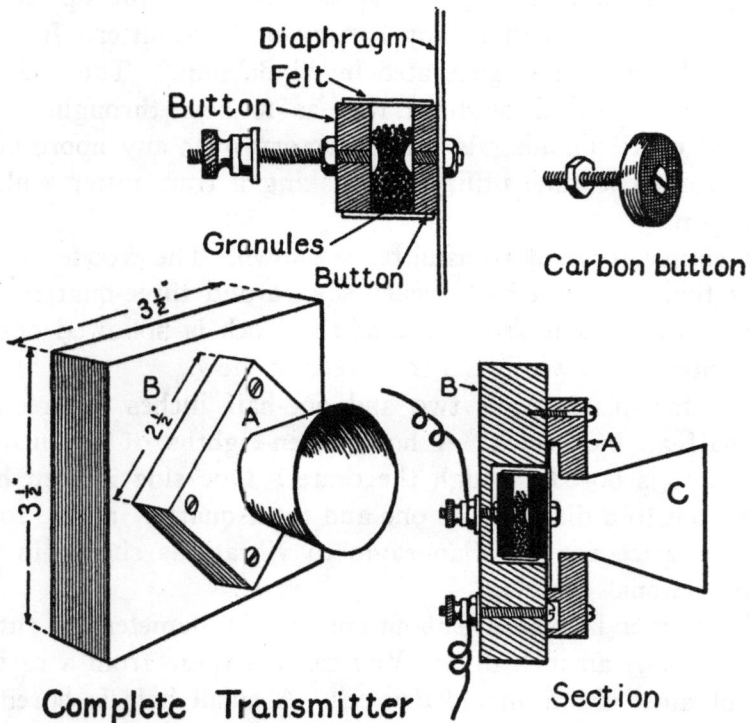

FIG. 145. — A Home-Made Telephone Transmitter.

of the carbon button mounted on the diaphragm, bind this strip with silk thread in such a manner that the strip forms a cylinder closed at one end with the button.

Fill the cylinder with polished carbon telephone transmitter granules to a depth of about one-eighth of an inch. These granules will have to be purchased from an electrical supply house. You can use carbon granules made by breaking up a battery carbon into fragments about as large as the head of a dressmaker's pin but they will not prove as satisfactory as transmitter carbon.

Slip a long machine-screw through the hole in the second carbon button and clamp it in place with a nut. Then place the carbon button in the cylinder so that it closes up the end. The space between the two buttons should be about three-sixteenths of an inch. Bind the flannel or felt around the button with a piece of silk thread so that it cannot slip out of place. The arrangement of the parts should now be the same as that shown by the cross-sectional drawing in the upper right-hand corner of Fig. 145.

The complete transmitter is assembled as shown in the same illustration.

A small tin funnel is fitted into the hole in the face-plate, A, to act as a mouthpiece.

A screw passes through the back, B, and connects to the diaphragm. The screw is marked "E" in the illustration. A binding-post is threaded on the screw so that a wire may be easily connected. The screw passing through the back carbon button also passes through a hole in the wooden back, and is clamped firmly in position with a brass nut so that the button is held very rigidly and cannot move. The front button, being attached to the diaphragm, is free to move back and forth with each vibration of the latter.

The carbon granules should fill the space between the buttons three-quarters full. They should lie loosely together, and not be packed in.

When connected to a battery and a telephone receiver the

current passes to the back button, through the mass of carbon granules into the front button and out. When the voice

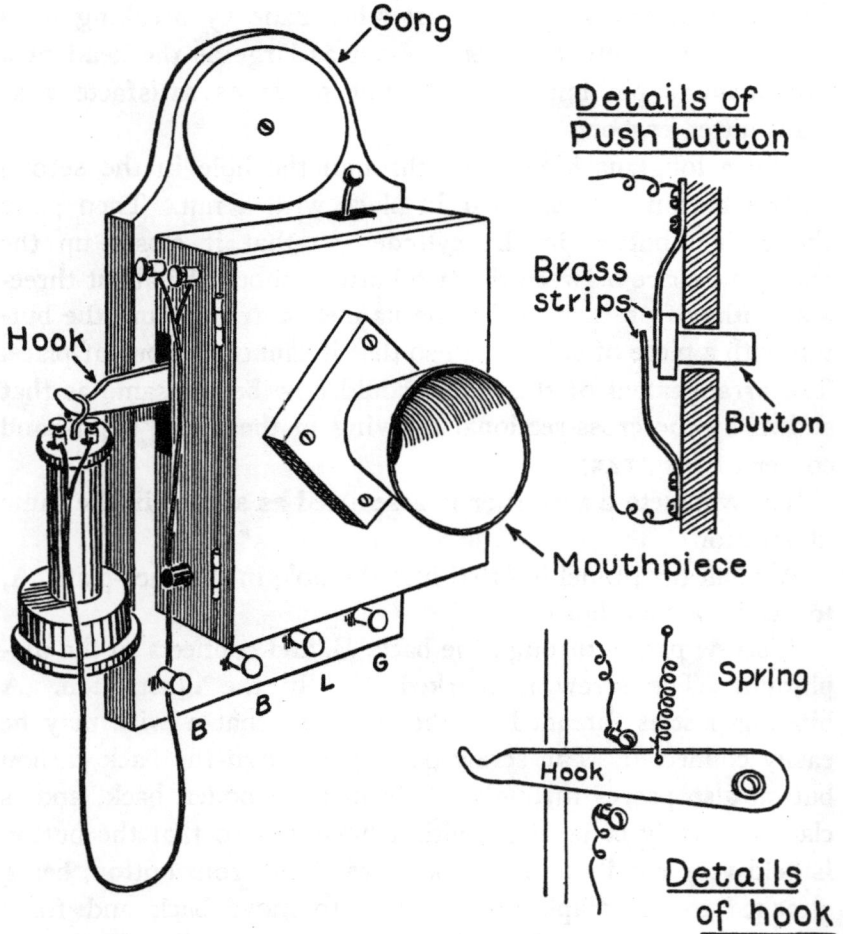

FIG. 146. — A Complete Telephone Instrument.

is directed into the mouthpiece, the sound waves strike the diaphragm and cause it to vibrate. The front button attached to it then also vibrates and constantly changes the pressure on the carbon granules. Each change in pressure is accompanied by an immediate change in resistance and consequently the amount of current flowing.

Figure 146 shows a complete telephone ready for mounting on the wall. It consists of a receiver, telephone transmitter, bell, hook and push-button. The bell is mounted on a flat base-board. The transmitter is similar to that just described, but is built into the front of a box-like cabinet. The box is fitted with a push-button at the lower right-hand corner. A simple method of making a suitable push-button is shown in the upper left-hand part of the illustration. It consists of two small brass strips arranged so that pushing a small wooden plug projecting through the side of the cabinet will bring the two strips together and make an electrical connection.

The "hook" consists of a strip of brass, pivoted at one end with a round-headed brass wood screw and provided with a small spring, so that when the receiver is taken off the hook it will move up and make contact with the screw, marked C in the illustration. When the receiver is on the hook, its weight will draw the latter down against the screw, D. The hook is mounted on the base-board of the telephone, and projects through a slot cut in the side of the cabinet.

Four binding-posts are mounted on the lower part of the base-board. The two marked B and B are for the battery. That marked L is for the "line," and G is for the ground connection or the return wire. The line-wire coming from the telephone at the other station enters through the binding-post marked L, and then connects to the hook. The lower contact on the hook is connected to one terminal of the bell. The other terminal of the bell leads to the binding-post marked G, which is connected to the ground, or to the second line-wire, where two are used.

The post, G, and one post, B, are connected together. The other post marked B connects to one terminal of the transmitter. The other terminal of the transmitter is connected to the telephone receiver. The other post of the telephone receiver leads to

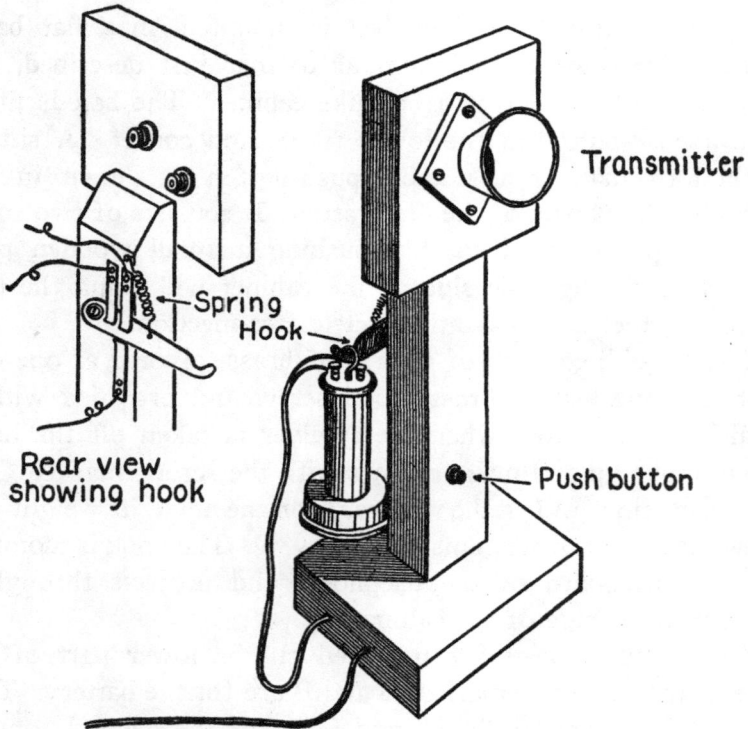

Transmitter

Spring
Hook

Rear view
showing hook

Push button

FIG. 147. — A Desk-Stand Type of Telephone.

the upper contact on the hook marked C. The push-button is connected directly across the terminals of the transmitter and the receiver so that when the button is pushed it short-circuits the transmitter and the receiver. When the receiver is on the hook and the latter is down so that it makes contact with D any current coming over the line-wire will pass through the bell and down through the ground or the return-wire to the other station, thus completing the circuit. If the current is strong enough it will ring the bell. When the receiver is lifted off the hook, the spring will cause the hook to rise and make contact with the screw marked C. This will connect the receiver, transmitter and the battery to the line so that it is possible to talk. If,

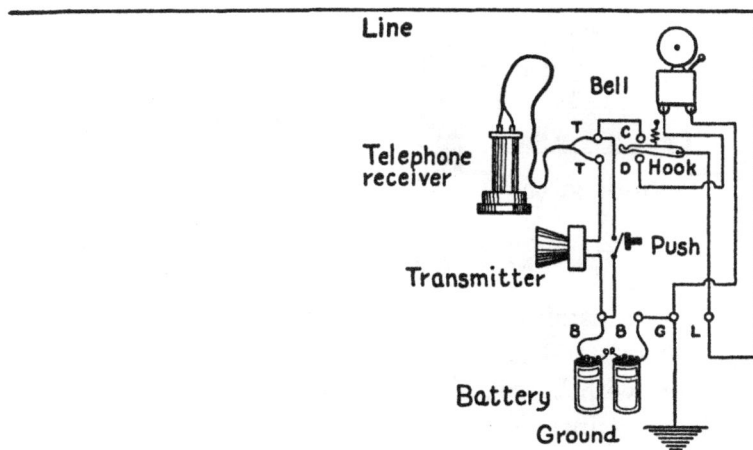

FIG. 148. — Diagram of Connection for the Telephone Instrument Shown in Figure 146.

however, it is desired to ring the bell on the instrument at the other end of the line, all that it is necessary to do is to press the push-button. This will short-circuit the receiver and the transmitter and ring the bell. The battery current is flowing over the line all the time when the receiver is up, but the transmitter and the receiver offer so much resistance to its flow that not enough current can pass to ring the bell until the resistance is cut out by short-circuiting them with the push-button.

The instruments at both ends of the line should be similar. In connecting them together care should be taken to see that the batteries at each end of the line are arranged so that they are in series and do not oppose each other. One side of the line may be a wire, but the return may be the ground, as already explained in the chapter on telegraph apparatus.

A transmitter of the "desk-stand" type may be made by

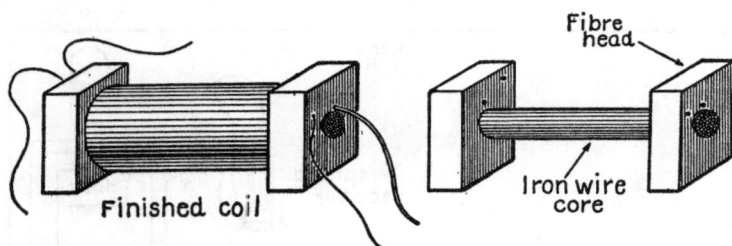

FIG. 149. — A Telephone Induction Coil.

mounting a transmitter upon an upright, provided with a base so that it may stand on a desk or a table.

It is fitted with a hook and a push-button, so that it is a complete telephone instrument with the exception of the bell and the battery. The battery and the bell may be located in another place and connected to the desk-stand by means of a flexible wire or "electrical cord."

Induction coils are used in telephone systems whenever it is necessary to work over a long distance. Such a system is more complicated, but is far superior to the system just described.

An induction coil consists of two fiber or hard-wood heads, about one inch square and one-quarter of an inch thick, mounted on the ends of an iron core composed of a bundle of small iron wires about two and one-half inches long. The core should be about five-sixteenths of an inch in diameter.

The core is covered with a layer of paper and then wound

with three layers of No. 22 B. & S. single-cotton-covered wire. These three layers of wire form the *primary*. The primary is covered with a layer of paper and then the secondary is wound on. The secondary consists of twelve layers

Only one Station is shown in the Diagram. The other is exactly similar. Observe that there are two Contacts operated by the Hook when the Receiver is lifted. Also that the Push-Button is a "two-way" push. It makes a contact with one point when at rest, and a contact with a second point when pushed. *P* and *P* are the *Primary* Wires of the Induction Coil, and *S* and *S* are the *Secondary*.

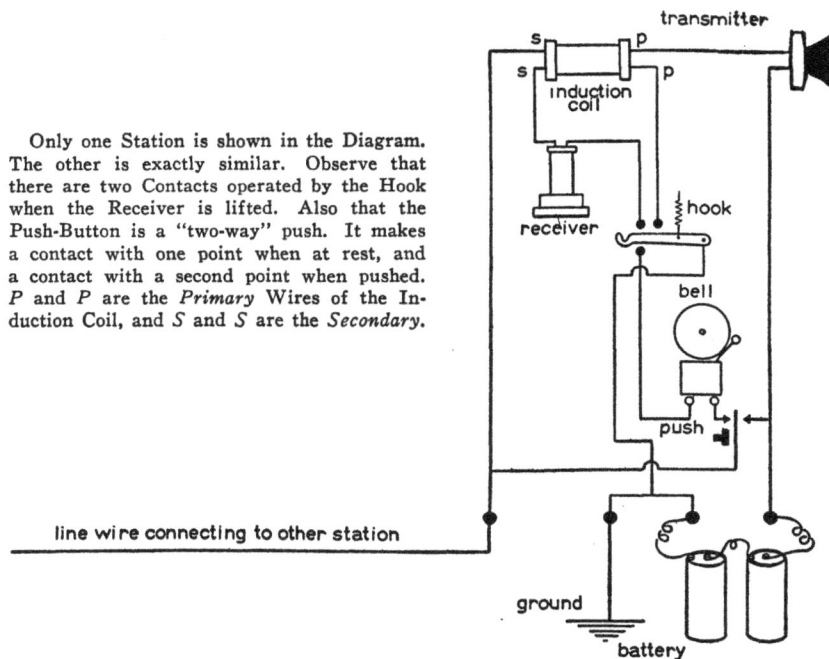

FIG. 150. — Diagram of Connection for a Telephone System Employing an Induction Coil at Each Station.

of No. 36 B. & S. single-silk-covered magnet wire. It is advisable to place a layer of paper between layers of the secondary winding, and to give each one a coating of shellac. The two secondary terminals of the coil are led out through holes in the fiber head and kept separate from the primary terminals.

The wiring diagram of a telephone system using an induction coil at each station is shown in Figure 150. The speech sent over a line using an induction coil system is much clearer and more easily understood than that on a line not using such a device.

In building telephone instruments or connecting them, care and accuracy will go a long way towards success. Telephony involves some very delicate and sensitive vibratory mechanical and electrical actions, and such instruments must be very carefully made.

CHAPTER XII

INDUCTION COILS

A Medical Coil or shocking coil, as it is properly termed, is nothing more or less than a small induction coil, and consists of a core, a primary winding, a secondary winding, and an interrupter. The principle of magnetic induction has already been explained in Chapter V. It might be well for the reader to turn back and reread those pages.

The human body possesses considerable resistance, and the voltage of one or two cells of battery is not sufficient to overcome that resistance and pass enough current through the body to be felt, except under exceptional conditions.

The simplest means employable for raising the voltage of a battery high enough to produce a shock is the induction coil.

The first step in making such a coil is to roll up a paper tube, five-sixteenths of an inch in diameter inside, and two and one-half inches long. The outer end of the paper is carefully glued, so that it will not unroll. The tube is filled with pieces of iron wire two and one-half inches long which have been straightened by rolling between two boards. The size of the iron wire may vary from No. 20 to No. 24 B. & S. gauge. Enough should be slipped into the tube to pack it tightly.

A square block, $1 \times 1 \times \frac{5}{16}$ inches, is cut out of fiber or a close-grained hard wood and a hole three-eighths of an inch in diameter bored through the center. One end of the tube containing the core is smeared with glue and slipped into the block. The end of the tube is allowed to project through about one-sixteenth of an inch. A second block, in the form of a circle three-quarters of an inch in diameter, one-quarter of an inch thick, and having a three-eighths of an inch hole through the center, is glued on the opposite end.

After the glue has dried, four small holes are drilled in the square head in the approximate positions shown in the illustration. The primary winding consists of four layers of No. 22

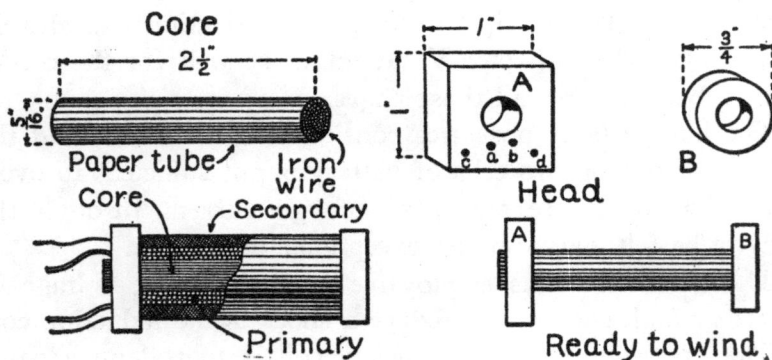

FIG. 151.—The Core and Windings of the Medical Coil.

B. & S. gauge magnet wire (it may be either silk or cotton, double or single covered) wound smoothly and carefully over the core. The terminals are led out of the holes a and b. The primary is covered with two layers of paper, and then enough secondary wound on to bring the total diameter of the coil to about eleven-sixteenths of an inch. The secondary wire must be much finer than the primary. It is possible to use any size from No. 32 to No. 36 B. & S. gauge and obtain good results. The insulation may be silk, cotton, or enamel.

The secondary terminals are led out through the holes c

and d. It is a wise plan to re-enforce these leads with a heavier piece of wire, because otherwise they are easily broken.

The interrupter is a simple arrangement capable of being improved upon. I have shown the simplest arrangement, so that all my readers will be able to build it, and those who want to improve it can do so.

If a small piece of silver is soldered to the spring and to the contact-point it will give better results. One terminal of the

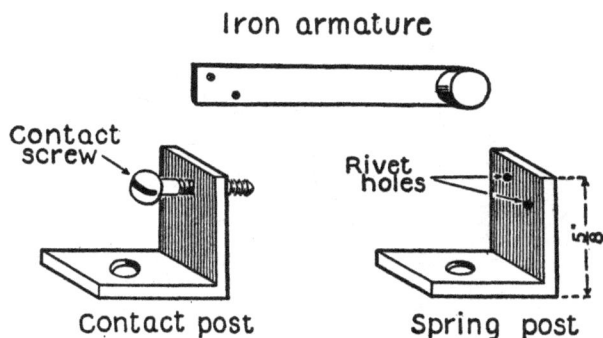

Iron armature

Contact screw

Rivet holes

Contact post

Spring post

FIG. 152. — Details of the Interrupter for the Medical Coil.

primary is connected to the interrupter spring and the other to a binding-post. The contact-post is also connected to a binding-post. If a battery is connected to the two binding-posts, the current will flow from one post through the coil to the interrupter spring, through the spring to the contact post, and thence back to the battery, making a complete circuit. As soon as the current flows, however, it produces magnetism which draws the spring away from the contact and breaks the circuit, cutting off the magnetic pull. The spring flies back to the contact but is drawn forward again immediately and repeats the operation continuously at a high rate of speed.

The secondary terminals are led out to two binding-posts to which are connected two electrodes or handles by means of flexible wires. The electrode may be made of two ordinary flat strips of sheet-metal or a piece of tubing. In the latter case, the wires may be connected by wedging them in with a cork. If the handles are grasped while the battery is connected to the primary posts and the interrupter is in operation a powerful shock will be felt. The shock may be regu-

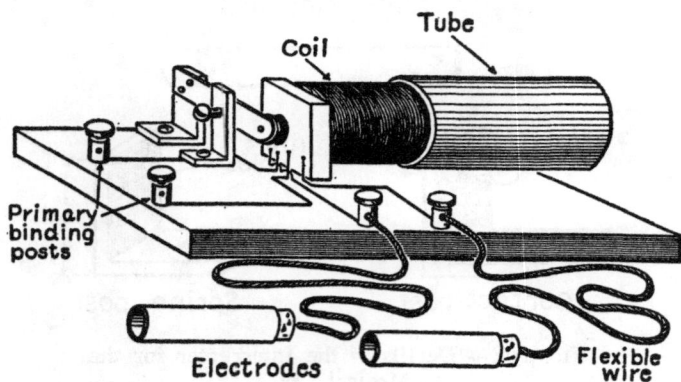

FIG. 153. — The Completed Medical Coil.

lated from a weak current that can hardly be felt to a very powerful one by providing the coil with a piece of iron tubing of about seven-eighths of an inch inner diameter and two inches long which will slip on and off the coil. When the tube is all the way on, the shock is very mild, and when all the way off, the shock is very strong. Of course any intermediate strength may be secured at stages between the two extremes.

The coil just described is harmless. It will give a strong shock, but the only result is to make the person receiving it drop the handles and not be anxious to try it again.

SPARK-COILS

A "spark-coil" is one of the most interesting pieces of apparatus an experimenter can possess. The experiments that may be performed with its aid are varied and many.

The purpose of a "spark-coil" is to generate high voltages which are able to send sparks across an air space that ordinary

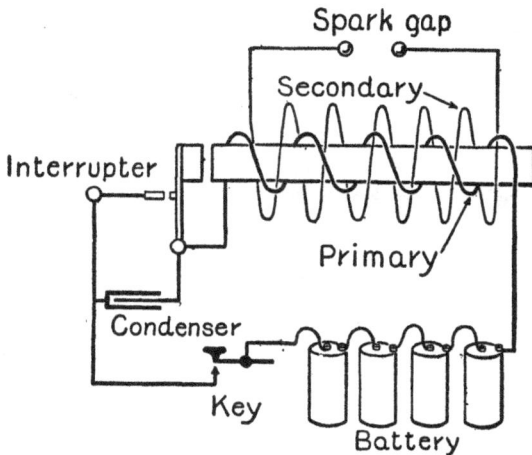

FIG. 154.— Diagram Showing the Circuit of a Spark-Coil.

currents of low voltage could not possibly pierce. The spark-coil is the same in principle as the small induction coils used as medical or shocking coils, but is made on a larger scale and is provided with a condenser connected across the terminals of the interrupter.

It consists of an iron core surrounded by a coil of heavy wire called the "primary," and by a second outside winding of wire known as the "secondary." The primary is connected to a few cells of battery in series with an interrupter. The interrupter makes and breaks the circuit, i. e., shuts the current on and off repeatedly.

Every time that the current is "made" or broken, a high voltage is induced in the secondary. By means of the condenser connected across the interrupter terminals, the current at "make" is caused to take a considerable fraction of time to grow, while at "break," the cessation is practically instantaneous. The currents induced in the secondary at break are so powerful that they leap across the space in a brilliant torrent of sparks.

Building a Spark-Coil

A spark-coil is not hard to construct, but it requires careful work and patience. It is not a job to be finished in a day,

Empty paper Tube

Tube filled with wire

FIG. 155. — Empty Paper Tube, and the Tube Filled with Core Wire Preparatory to Winding on the Primary.

but time must be liberally expended in its construction. Satisfactory results are easily obtained by any one of ordinary mechanical ability if patience and care are used.

For the benefit of those who wish to build a larger coil than the one described in the following text, a table showing the dimensions of two other sizes will be found.

The core is made of soft iron wires * about No. 20 or 22
B. & S. gauge, cut to exact length. Each piece should be six
inches long. The wire must be annealed before it can be used.
This is accomplished by tying the wire in a compact bundle
and placing it in a wood fire where it will grow red-hot. When

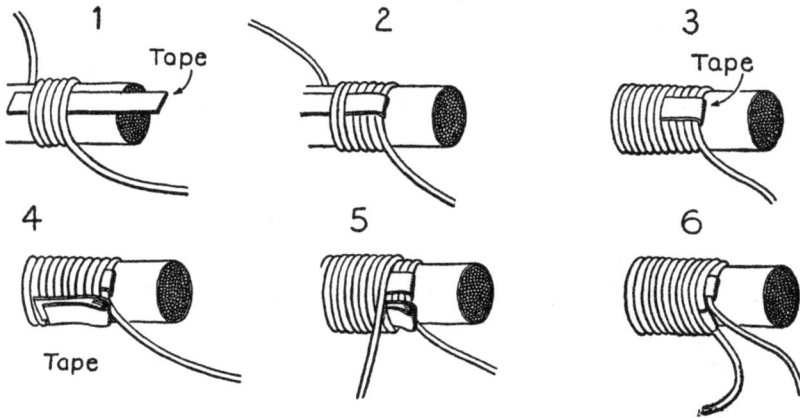

FIG. 156. — The Various Steps in Winding the Primary on the Core and
Fastening the Ends of the Wire.

this stage is reached, cover the wire with ashes and allow the
fire to die away.

Cut a piece of tough wrapping paper into strips six inches
long and about five inches wide. Wrap it around a stick or
metal rod one-half of an inch in diameter, so as to form a tube
six inches long and having a diameter of one-half of an inch.
Glue the inside and outside edges of the paper so that the tube
cannot unroll and then slip it off the stick. Fill the tube with
the six inch wires until it is packed tightly and no more can
be slipped in.

The primary consists of two layers of No. 18 B. & S. gauge

* The black iron wire known as stove-pipe wire.

cotton-covered wire wound over the core for a distance of five inches. One-half pound of wire is more than enough for one primary. The wire must be wound on very smoothly and carefully. In order to fasten the inside end so that it will not become loose, place a short piece of tape lengthwise of the core and wind on two or three turns over it. Then double the end back and complete the winding. After the first layer is finished, give it a coat of shellac and wind on the second layer. The end of the wire is wound with a piece of tape and

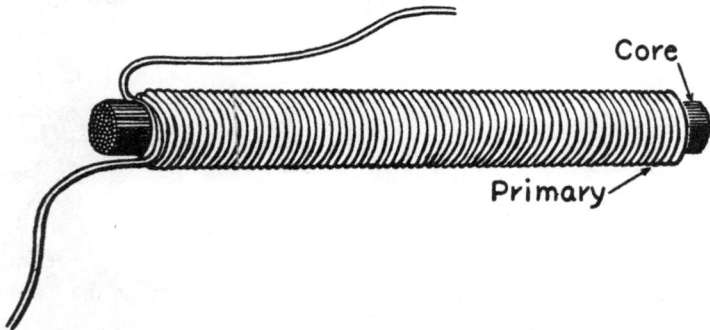

FIG. 157. — The Core and Completed Primary Winding.

fastened by slipping through a loop of tape embedded under the last few turns. The illustrations will explain more clearly just how this is accomplished. The second layer is then given a coat of shellac and allowed to dry. After it is dry, wrap about fifteen layers of paper which have been soaked in paraffin around the primary.* This operation should be performed in a warm place, over a fire or lighted lamp where the paraffin may be kept soft, so that the paper will go on tightly.

The coil is now ready to receive the secondary winding.

* The best material to use for this purpose is "Empire" paper or cloth. It is not necessary to use paraffin with Empire paper or cloth.

The core and primary which have been described are suitable for a secondary giving sparks from one-half to three-fourths of an inch long.

The secondary winding consists of several thousand turns of very fine wire wound on in smooth, even layers with paper between each two layers.

The following table shows the size and amount of wire required. The wire may be either enamel, or silk insulated. Enamel covered wire is preferred.

Size of Coil	Size of Wire	Amount
½ inch	36 B. & S.	10 ounces
1 inch	34 B. & S.	1 lb.
1½ inch	34 B. & S.	2 lbs.

The means for supporting and turning the coil in order to wind on the secondary may be left somewhat to the ingenuity

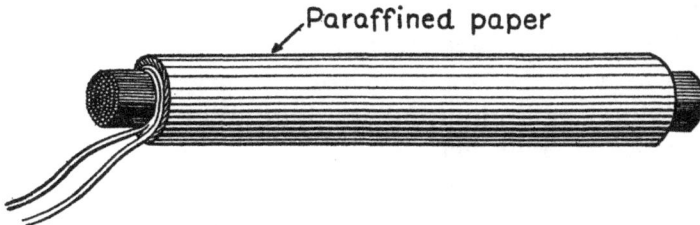

FIG. 158. — The Primary Winding Wrapped with Several Layers of Paraffined Paper Ready for the Secondary Winding.

of the young experimenter. The following suggestion, however, is one which experience has proved to be well worth following out, and may be applied to other things than the construction of an induction coil. It is quite natural for most boys, for some reason or other, to try to avoid spending time and labor on anything which will aid them in their work. They are always in such a hurry and so anxious to see something completed that they direct all their energy to that end rather than spend part of their time in constructing some little device which would really lighten the other work and go a long way towards insuring its successful completion.

I am placing particular stress on the care required in winding a spark-coil because if such suggestion is ignored in the anxious endeavor to finish the coil as soon as possible, in every such instance the coil will be a failure.

The illustration shows a simple form of winder, with which the operation of winding the secondary is a very slow one, but, on the other hand, it is possible to do very accurate, careful winding with the aid of such a device. The parts may all be made from wood.

The "chucks" fit tightly over the ends of the core so that when the handle is turned, the coil will revolve also. The spring serves to keep the chucks snugly against the coil ends, so that they will not slip.

From one-half to five-eighths of a pound of wire will be required to wind the secondary coil. A large number of strips of thin waxed paper must be cut five inches wide. The inside terminal, or "beginning" end of the wire is tied around the insulating tube near the left-hand end. The spool of wire must be placed in a position where it will revolve freely without strain on the wire. No. 36 is very fine and easily broken, so use the utmost care to guard against this mishap.

Wind on a smooth, even layer of wire, permitting each turn to touch the other, but none to lap over. Carry the winding to within three-quarters of an inch of the ends of the insulating tube and then wind on two layers of the waxed paper. The paper must be put on smoothly and evenly, so as to afford a firm foundation for the next layer. The wire is wrapped around with the paper, so that the next layer starts three-quarters of an inch from the edge. A second layer is then wound on very carefully, stopping when it comes one-half inch from the edge. Two more layers of paper are put on, and the process repeated, alternately winding on paper and wire until the stated quantity of the latter has been used up.

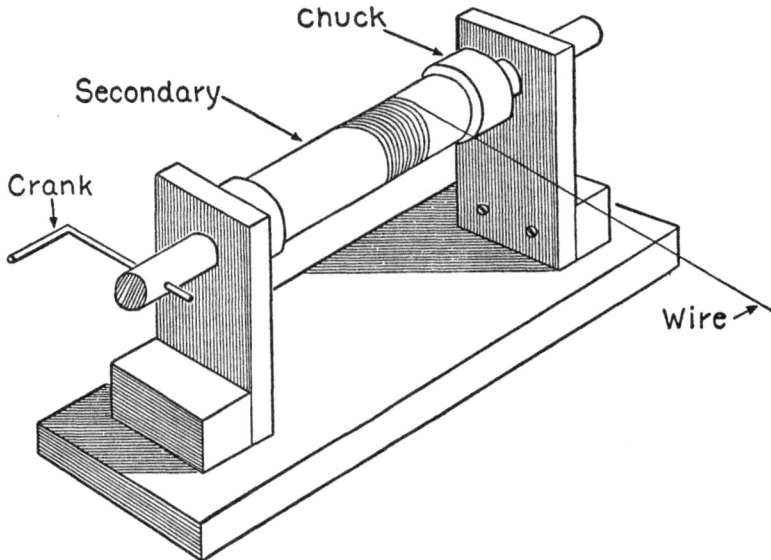

FIG. 159. — A Simple Device for Winding the Secondary.

In winding the coil, remember that if at any point you allow the winding to become irregular or uneven, the irregularity will be most exaggerated on the succeeding layers. For this reason, do not allow any to occur. If the wire tends to go on unevenly, wrap an extra layer of thick paper around underneath so as to offer a smooth foundation, and you will find the difficulty remedied.

An efficient interrupter for a coil cannot be easily made, and it is best to buy one at an automobile accessory store. Ask for a set of interrupter parts for a Model "T" Ford coil. The interrupter will play a very important part in the successful working of the coil, and its arrangement and construction are important. Interrupters like that shown in the illustration and used for automobile ignition, will be found the best.

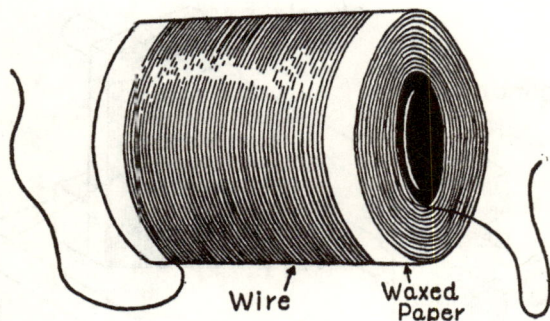

Fig. 160. — Completed Secondary Winding.

You can also purchase a condenser ready made. The condenser for any automobile ignition coil will serve. The condenser may be home-made. It consists of alternate sheets of tinfoil and paraffined paper, arranged in a pile as shown in the illustration. The following table gives the proper sizes for condensers for three different coils.

Size of Spark-Coil	No. Sheets	Tinfoil Size of Sheets
½ inch	50	2 × 2
1 inch	100	7 × 5
1½ inch	100	8 × 6

The paper must be about one-half inch larger all the way around, so as to leave a good margin. The alternate sheets of tinfoil, that is, all on one side and all on the other, are connected. The condenser is connected directly across the terminals of the interrupter.

There are various methods of mounting a coil, the most common being to place it in a box with the interrupter at one end. Perhaps, however, one of the neatest and also the simplest methods is to mount it in the manner shown in the illustration.

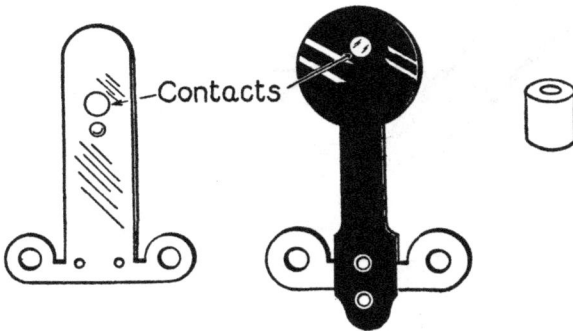

FIG. 161. — Parts of the Interrupter.

The end-pieces are cut out of wood. No specific dimensions can be given, because the diameter of the coils will vary somewhat according to who winds them and how tightly they are made. The coil is enclosed in a tube made by rolling up a strip of cardboard and then giving it a coat of shellac. The tube may be covered by a strip of black cloth, so as to improve its appearance.

The vibrator is mounted on the end. The core projects through a hole in the wood near the end of the vibrator spring so that the latter will be drawn in by the magnetism of the core when the current flows. The condenser may be placed in the hollow box which forms the base of the coil.

The secondary terminals of the coil are mounted on a small strip of wood bridging the two coil ends.

One terminal of the primary is connected to a binding-post mounted on the base, and the other led to the vibrator spring. The vibrator yoke is connected to a second binding-post on the base. One terminal of the condenser is connected to the spring, and the other to the yoke.

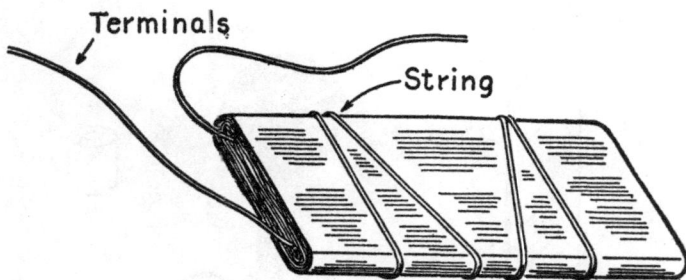

FIG. 162. — The Condenser.

Four cells of dry battery should be sufficient to run the coil and cause it to give a good one-half-inch spark if built according to the directions here given. The interrupter will require adjusting and a position of the adjusting screw will soon be found where the coil works best.

EXPERIMENTS WITH A SPARK-COIL *

Electrical Hands. Many extraordinary and interesting experiments may be performed with the aid of a spark-coil.

The following experiment never fails to amuse a party of friends, and is mystifying and weird to the ordinary person, unacquainted with the secret of its operation.

* Those experimenters who may not wish to build a spark-coil can perform most of the experiments with the spark-coil from a Model T Ford. These are inexpensive.

The primary of an ordinary spark-coil is connected in series with a twelve-volt battery and telephone transmitter. A small switch is included in the circuit to break the current and prevent needless waste of the battery when the apparatus is not in immediate use. The secondary terminals of the induction coil are led by means of an insulated wire to the adjoining room

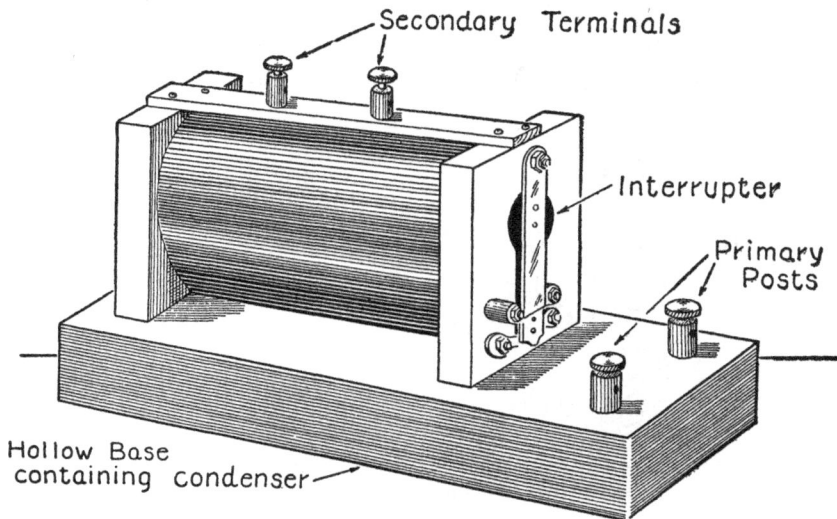

FIG. 163. — The Completed Coil.

where they terminate in a pair of scissors, or some other small metallic object which may be clasped in the hand.

Each of two persons, wearing dry shoes or rubber-soled slippers, grasps the terminal of one wire in one hand. The other hand is placed flat against the ear of a third person, with a piece of dry linen paper intervening between the hands and the head. If a fourth person, in the room where the induction coil is located, then closes the small switch and speaks into the telephone transmitter, the person against whose ears the hands are being held will hear the speech very distinctly. The ticking of a watch held against the mouthpiece of the transmitter will be heard with startling clearness.

The principle governing the operation of the apparatus is very simple. Almost every experimenter is familiar with the ordinary electrical condenser, which consists of alternate sheets of paraffined paper and tinfoil. When this is connected to a source of electricity of high potential, but not enough so as to puncture the paper dielectric, the alternate sheets of tinfoil

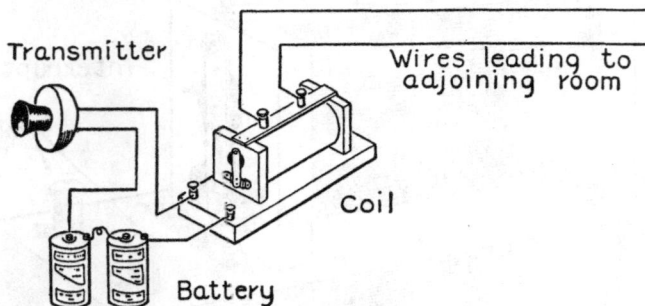

FIG. 164. — Diagram Showing How To Connect the Apparatus for the "Electrical Hands" Experiment.

will become oppositely charged and attract each other. If the circuit is then broken the sheets will lose their charge and also their attraction for one another. If the tinfoil sheets and paper are not pressed tightly together, there will be a slight movement of the tinfoil and paper which will correspond in frequency to any fluctuations of the charging current which may take place.

The head of the third person and the hands held against his head act like three tinfoil sheets of a condenser, separated by two sheets of paper. The words spoken in the transmitter cause the current to fluctuate and the induction coil raises the

potential of the current sufficiently to charge the condenser
and cause a slight vibration of the paper dielectric. The vibrations correspond in strength and speed to those of the

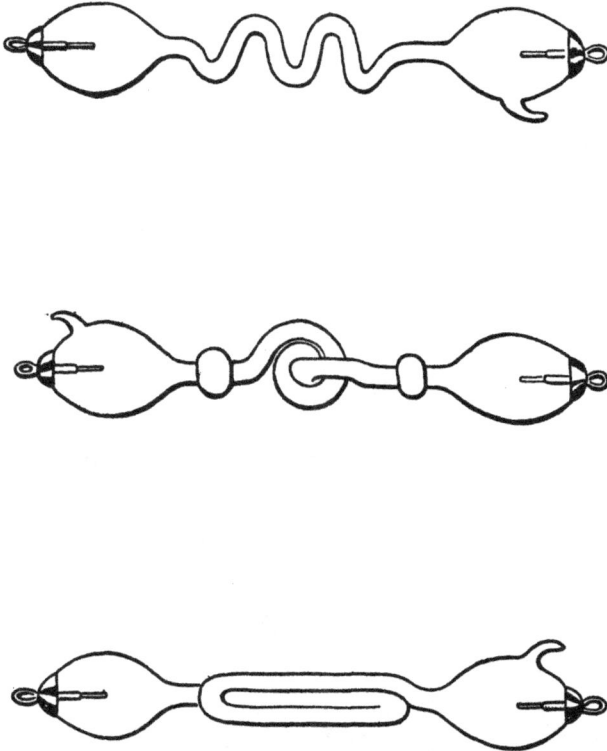

FIG. 165. — Geissler Tubes.

voice, and so the words spoken in the transmitter are audible
to the person over whose ears the paper is pressed.

Everything about the apparatus must be as dry as possible, to insure its successful operation. The people holding the
wires in their hands should stand on a carpeted floor. Always
be very careful to tighten the adjusting screw and block the

interrupter on the coil, so that by no means can it possibly commence to operate, or the person listening, instead of "hearing things," will become the victim of a rather painful, practical joke.

GEISSLER TUBES

The most beautiful and surprising effects may be obtained by lighting Geissler tubes with a coil. The tubes are made in in-

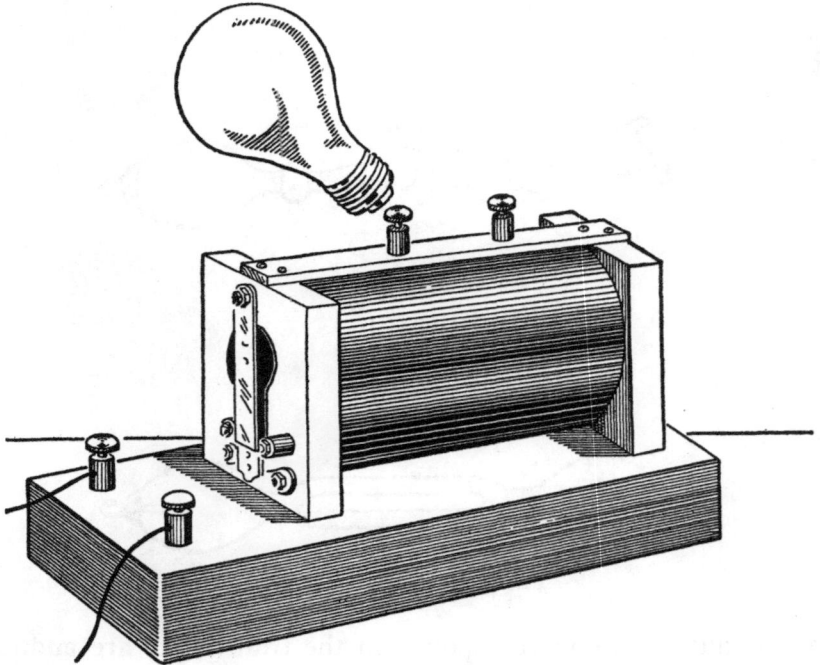

FIG. 166. — The Ghost Light Experiment.

tricate and varied patterns of special glass, containing fluorescent minerals and salts, and are filled with different rarefied gases. When the tubes are connected to the secondary of a spark-coil by means of a wire fastened to the little rings at the end, and the coil is set in operation, they light up in the most wonderful way imaginable. The rarefied gases and minerals in the glass throw out beautiful iridescent colors, lighting up a dark room with a weird flickering light. Every tube is usually of a different pattern and has a combination of different colors. The most beautiful tubes are those provided with a double wall containing a fluorescent liquid, which heightens the color effects when the tube is lighted.

Eight to ten tubes may be lighted at once on an ordinary coil by connecting them in series.

GHOST LIGHT

If you grasp the bulb of an incandescent electric lamp in one hand and touch the base to one side of the secondary when the coil is in operation, the bulb will emit a peculiar violet light in the dark.

PUNCTURING PAPER

If you place a piece of heavy paper or cardboard between two sharp wires connected to the secondary of a spark-coil and start the coil working, the paper will be pierced.

A PRACTICAL JOKE

This action of the coil may be made the basis of an amusing joke. Offer a friend who smokes a cigarette which has been prepared in the following way.

Place the cigarette on a piece of sheet-metal which is connected to one side of the secondary. By means of an insulated handle so that you will not get a shock, move the other wire above the cigarette, allowing an occasional thin, straggling spark to jump. The paper will be pierced with numerous holes which are so fine that they can hardly be seen.

If your friend tries to light the cigarette he will waste a box of matches without being able to get one good puff, because the little invisible holes in the paper will spoil the draft.

An Electric Garbage-Can

If there are any dogs in your neighborhood that have a habit of extracting things from your garbage can, place the latter

Fig. 167. — An Electrified Garbage-Can.

on a piece of dry wood. Lead a well insulated wire from one secondary terminal of your coil to the can. Ground the other secondary terminal. If you see a dog put his nose in the can press your key and start the coil working. It will not hurt the dog, but he will get the surprise of his life. He will go for home as fast as he can travel and will not touch that particular can again, even if it should contain some of the choicest canine delicacies.

Photographing an Electric Discharge

The following experiment must be conducted in a dark room with the aid of a ruby photographic lamp, as otherwise the plates used would become lightstruck and spoiled.

Place an ordinary photographic plate or film on a piece of sheet-metal with the coated side of the plate upwards. Connect one of the secondary terminals of the spark-coil to the piece of sheet-metal.

Then sift a thin film of dry starch powder, sulphur, or talcum through a piece of fine gauze on the plate. Lead a sharp-pointed wire from the other secondary terminal of the coil to the center of the plate and then push the key just long enough to make one spark.

Wipe the powder off the plate and develop it in the usual manner of films and plates. If you cannot do developing yourself, place the plate back in its box and send it to some friend, or to a photographer.

The result will be a negative showing a peculiar electric discharge, somewhat like sea-moss in appearance. No two such photographs will be alike and the greatest variety of new designs, etc., imaginable may be produced in this manner.

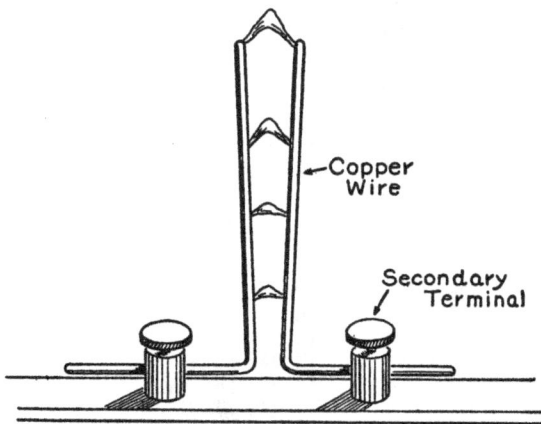

FIG. 168. — Jacob's Ladder.

JACOB'S LADDER

Take two pieces of bare copper wire about eight inches long and bend them at right angles. Place them in the secondary terminals of a spark-coil. Bend them so that the vertical portions are about one-half of an inch apart at the bottom and one inch apart at the top. Start the coil working, and the sparks will run up the wires from the bottom to the top and appear very much like the rungs in a ladder.

X-RAYS

Most young experimenters are unaware what a wonderful and interesting field is open to the possessor of a small X-ray

Anodes

Cathode

FIG. 169.—A Small X-ray Tube.

tube. It is possible to obtain small X-ray tubes which will operate satisfactorily on an inch and one-half spark-coil. They usually cost about four dollars and a half. With such a tube and a fluoroscope it is possible to see the bones in the human hand, the contents of a closed purse, etc.

The tube is made of glass and contains a very high vacuum. The long end of the tube contains an electrode called the *cathode*. The short end contains two electrodes called *anodes,* one perpendicular to the tube and the other diagonal.

The tube is usually clamped in a wooden holder called an X-ray tube stand. The tube should be so adjusted that the X-rays which are reflected from the diagonal anode will pass off in the direction shown by the dotted lines in Figure 172.

The fluoroscope is a cone-shaped wooden box fitted with a screen composed of a sheet of paper covered with crystals of a chemical called platinum-barium-cyanide.

The opposite end of the box is fitted with a covering of felt or velvet which shuts off the light around the eyes and nose when you look into the fluoroscope and hold it tightly against the face.

A fluoroscope may be purchased complete, or the platinum-barium-cyanide screen purchased separately and mounted on a box as shown in the illustration.

The two anodes of the tube should be connected and led to one terminal of a spark-coil capable of giving a spark at least one and one-half inches long. Another wire should be led from the cathode of the tube to the other terminal of the coil.

When it is desired to inspect any object, such as the hand, it must be held close to the screen of the fluoroscope and placed between the latter and the tube in the path of the X-rays. The X-rays are thrown forth from the tube at an angle of 45 degrees from the diagonal anode.

Look into the fluoroscope and it should appear to be filled with a green light. If not, the battery terminals connected to the primary of the coil should be reversed, so as to send the current through in the opposite direction.

The X-rays will cause the chemicals on the screen to fluoresce and give forth a peculiar green light. If the hand is held against the screen, between the screen and the tube, the X-rays

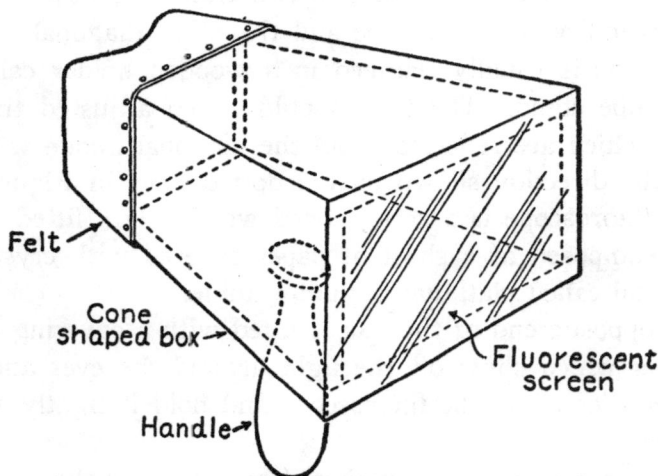

FIG. 170. — The Fluoroscope

will pass through the hand and cast a shadow on the screen. They do not pass through the bones as easily as they do through the flesh and so will cast a shadow of the bones in the hand on the screen, and if you look closely you will be able to see the various joints, etc.

The interrupter on the coil should be carefully adjusted so that the light does not flicker too much.

If it is desired to take X-ray pictures, a fluoroscope is unnecessary.

Turn the tube around so that the X-rays point downward.

Shut the battery current off so that the tube is not in operation until everything else is ready.

Place an ordinary photographic plate, contained in an ordinary plate-holder, directly under the tube with the gelatin side of the plate upwards.

Place the hand flat on the plate and lower the tube until it is only about three inches above the hand. Then start the coil

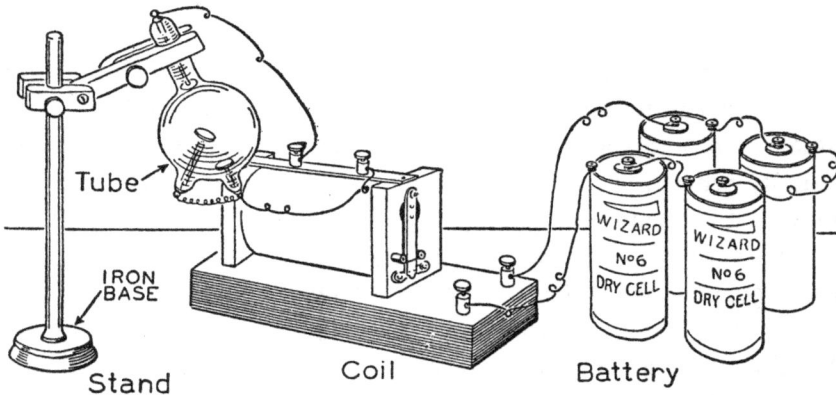

FIG. 171. — The Complete X-Ray Outfit.

working so that the tube lights up and permit it to run for about fifteen minutes without removing the hand. Then turn the current off and develop the plate in a dark room.

It is possible to obtain a very good X-ray photograph of the hand in this manner. Photographs showing the skeleton of a mouse, nails in a board, coins in a purse, a bullet in a piece of wood, etc., are a few of the other objects which make interesting pictures.

CHAPTER XIII

TRANSFORMERS

In most towns and cities where electricity for light and power is carried over long distances, it will be noticed that small iron boxes are fastened to the poles at frequent intervals, usually wherever there is a group of houses or buildings supplied with the current. Many boys know that the boxes contain "transformers," but do not always understand exactly what their purpose is, and how they are constructed.

When it is desired to convey electrical energy to a distance, for the purpose of producing either light or power, one of the chief problems to be faced is, how to reduce to a minimum any possible waste or loss of energy during its transmission. Furthermore, since wires and cables of large size are very costly, it is desirable that they be as small as possible and yet still be able to carry the current without undue losses.

It has already been explained that wires offer resistance to an electrical current, and that some of the energy is lost in passing through a wire because of this resistance. Small wires possess more resistance than large ones, and if small wires are to be used, in order to save on the cost of the transmission line, the loss of energy will be greater, necessitating some method of partially reducing or overcoming this fault.

In order to explain clearly how the problem is solved, the electric current may for the moment be compared to a stream of water flowing through a pipe.

The illustration shows two pipes, a small one and a large one, each supposed to be connected to the same tank, so that the pressure in each is equal, and it is clearly apparent that more water will flow out of the large one than out of the small one. If ten gallons of water flow out of the large pipe in one minute, it may be possible that the comparative sizes of the pipes are such that only one gallon of water will flow out of the small one in the same length of time.

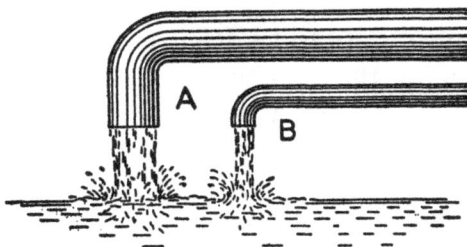

Fig. 172. — Comparison Between an Electric Current and Flow of Water.

But in case it should be necessary or desirable to get ten gallons of water a minute out of a small pipe such as B, what could be done to accomplish it?

The pressure could be increased. The water would then be able better to overcome the resistance of the small pipe.

This is exactly what is done in the distribution of electric currents for power and lighting. The pressure or potential is increased to a value where it can overcome the resistance of the small wires.

But unfortunately it rarely happens that electrical power can be utilized at high pressure for ordinary purposes. For instance, 110 volts is usually the maximum pressure required by incandescent lamps, whereas the pressure on the line wires issuing from the power-house is generally 2,200 volts or more.

Such a high voltage is hard to insulate, would kill most people coming into contact with the lines, and is otherwise dangerous.

Before the current enters a house, therefore, some apparatus is necessary, which is capable of reducing this high pressure to a value where it may be safely employed.

This is the duty performed by the "transformer" enclosed in the black iron box fastened on the top of the electric light poles about the streets.

If a transformer were to be defined it might be said to be a device for changing the voltage and current of an *alternating* circuit in pressure and amount.

The word, alternating, has been placed in italics because it is only upon alternating currents that a transformer may be successfully employed. Therein, also, lies the reason why alternating current is supplied in some cases instead of direct current. It makes possible the use of transformers for lowering the voltage at the point of service.

Many boys possessing electrical toys and apparatus operating upon direct current only, have bemoaned the fact that the lighting system furnished alternating current. One power-house usually furnishes the current for several communities and the energy has to be carried a considerable distance. So alternating current is employed.

The Difference Between Alternating and Direct Currents

A direct current is one which passes in one direction only. It may be represented by a straight line.

An alternating current is one which reverses its direction and passes first one way and then the other. It may be represented by a curved line, shown in Figure 173. It starts at *zero* and gradually grows stronger and stronger. Then it commences to die away until no current is flowing. At this point it reverses and commences to flow in the opposite direction, rising gradually and then dying away again.

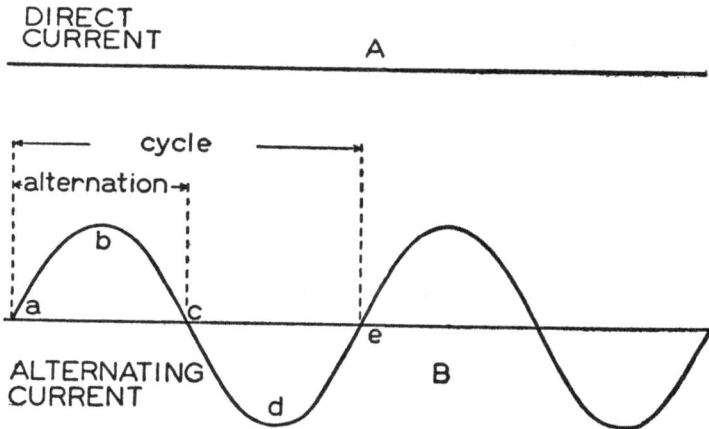

FIG. 173. — Pictorial Representation of a Direct and an Alternating Current.

This is repeated a definite number of times per second; when the current rises from zero, reverses and returns to zero, it is said to pass through a *cycle*.

The part of the curved line from *a* to *b* represents the first part of the current, when it is rising. From *b* to *c* represents its fall. The point at which the curved line crosses the straight line is zero. At *c* the current crosses the line and commences to flow in the opposite direction until it reaches *d*, at which point it dies away and again crosses the line to flow in its original direction and *repeat the cycle*.

In electrical parlance, that part of the current from *a* to *c* or from *c* to *e* is known as an *alternation*. From *a* to *e* is called a cycle.

The reason why alternating current is often used in place of direct current is that it can be sent over the wires for long distances more economically than direct current. This is more fully explained farther on in the chapter dealing with a step-down transformer.

The number of *cycles* taking place in one second is known as the *frequency* of the current. The usual *frequency* of commercial alternating currents is 60 cycles per second or 7200 alternations per minute.

THE ALTERNATING CURRENT SYSTEM FOR LIGHT AND POWER

The large dynamos located at the power-house generate alternating current. The alternating current passes into a "step-up" transformer which raises the potential to 2,200 volts or more. It is then possible to use much smaller line wires, and to transmit the energy with smaller loss than if the current were sent out at the ordinary dynamo voltage. The current passes over the wires at this high voltage, but wherever connection is established with a house or other building, the "service" wires which supply the house are not connected directly to the line wires, but to a "step-down" transformer which lowers the potential of the current flowing into the house to about 110 volts.

In larger cities where the demand for current in a given area is much greater than that in a small town, a somewhat different method of distributing the energy is employed.

The alternating current generated by the huge dynamos at the "central" station is passed into a set of transformers which in some cases raise the potential as high as five or six thousand volts. The current is then sent out over cables or "feeders" to various "sub" stations, or "converter" stations, located in various parts of the city. Here the current is sent through a set of "step-down" transformers which reduce the potential.

A transformer in its simplest form consists of two independent coils of wire wound upon an iron ring. When an alternating current is passed through one of the coils, known

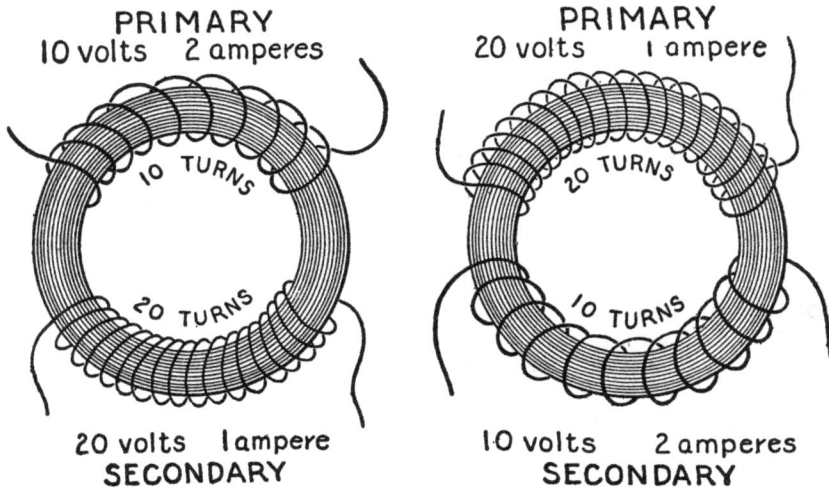

FIG. 174. — Diagrams Showing the Principle of the Step-Up and Step-Down Transformers.

as the primary, it produces a magnetic field which induces a current of electricity in the other, or secondary, coil.

The potential or voltage of the current in the secondary is in nearly the same ratio to the potential of the current passed into the primary as the number of turns in the secondary is to the number of turns in the primary.

Knowing this, it is very easy to arrange a transformer to "step" the potential up or down as desired. One of the transformers in Figure 174 represents a "step-up" transformer having ten turns of wire on the primary and twenty turns on the secondary. If an alternating current of 10 volts and 2 amperes is passed into the primary, the secondary winding will double the potential, since it has twice as many turns as the primary and the current delivered by the secondary will be approximately 20 volts and 1 ampere.

The action may be very easily reversed and a "step-down" transformer arranged by placing twenty turns of wire on the primary and ten turns on the secondary. If a current of 20 volts and 1 ampere is passed into the primary, the secondary will deliver a current of only 10 volts and 2 amperes, since it contains only half as many turns.

A circular ring of iron wire wound with two coils would in many respects be somewhat difficult to construct, and so the iron core is usually built in the form of a hollow rectangle and formed of sheets of iron.

How to Build a Step-Down Transformer

It is often desirable to have at hand an alternating current of low voltage for experimental purposes. Such a current may be used for operating induction coils, motors, lamps, toy railways, etc., and is quite as satisfactory as direct current for many purposes, with the possible exception of electro-plating and storage-battery charging, for which it cannot be used without being rectified. When the supply is drawn from the 110-volt lighting circuit and passed through a small "step-down" transformer, the alternating current is not only cheaper but more convenient. A transformer of about 100 watts capacity, capable of delivering a current of 10 volts and 10 amperes from the secondary will only draw little more than one ampere from the 110-volt circuit. This current is equal only to that consumed by three ordinary 40 watt lamps, making it possible to operate the transformer to its full capacity for about one cent an hour. A further advantage is the fact that a "step-down" transformer enables the small boy to use the lighting current for operating electrical toys without danger of receiving a shock.

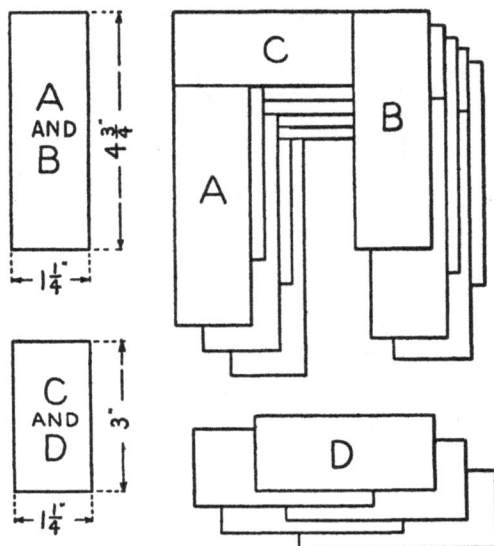

FIG. 175. — Details of the Core for a Small Step-Down Transformer.

The transformer described in the following pages can be easily built. It should make a valuable addition to an electrical laboratory, provided the directions are carefully followed and pains are taken to make the insulation perfect.

The capacity of the transformer is approximately 100 watts. The dimensions and details of construction described and illustrated are those of a transformer intended for use upon a lighting current of 110 volts and 60-cycles frequency. The frequency of most alternating current systems is 60 cycles.

The first part to be considered in the construction of a transformer is the core. The core is made up of thin sheet-iron strips. The iron may be secured from a hardware store or plumbing shop by ordering "stove-pipe iron." Have the iron cut into strips 1¼ inches wide and 24 inches long. Then, using a pair of tinner's shears, cut the long strips into pieces 3 inches

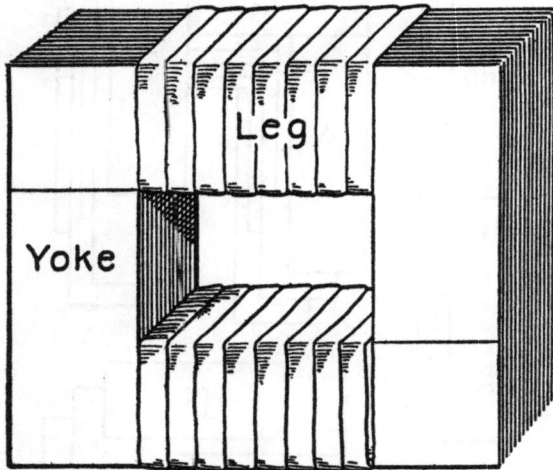

FIG. 176. — The Core, Assembled and Taped
Ready for Winding.

and 4¾ inches long until you have enough to make a pile of each 2½ inches high when they are stacked up neatly and compressed. The long strips are used to form the "legs" of the core, and the short ones the "yokes."

The strips are assembled according to the diagram. The alternate ends overlap and form a hollow rectangle 4¼ ×6 inches. The core should be pressed tightly together and the legs bound with three or four layers of insulating tape preparatory to winding on the primary. After the legs are bound, the yoke pieces may be pulled out, leaving the legs intact.

Four fiber heads, 2½ inches square and ⅛ of an inch thick are made. A square hole 1¼ × 1¼ inches is cut in the center. Two of these are placed on each of the assembled legs.

The primary winding consists of one thousand turns of No. 20 B. & S. gauge single-cotton-covered magnet wire. Five hundred turns are wound on each leg of the transformer. The wire should be wound on very smoothly.

The two legs should be connected in series. The terminals are protected and insulated by covering with some insulating tape rolled up in the form of a tube.

The secondary winding consists of one hundred and twenty turns of No. 14 B. & S. gauge double-cotton-covered wire. Sixty turns are wound on each leg, over the primary, several layers of paper being placed between the two.

A "tap" is brought out at the 24th, 48th, 72nd, 96th, 102nd, 103rd and 114th turn. The taps are made by soldering a narrow strip of sheet-copper to the wire at proper intervals. Care must be taken to insulate each joint and tape with a small strip of insulating tape so that there is no danger of a short circuit being formed between adjacent turns.

After the winding is completed the transformer is ready for assembling. The yoke pieces of the core should be slipped into position and the hole carefully lined up. The transformer itself is now ready for mounting.

The base-board measures 11 × 7¾ × ⅞ inches.

The transformer rests upon two wooden strips, *A* and *B,* 4¼ inches long, 1¼ inches wide, and ¾ of an inch high. The strips are nailed to the base so that they will come under the ends of the core outside of the fiber heads.

The transformer is held to the base by two tie-rods passing through a strip, *C,* 6 inches long, one-half of an inch thick and three-quarters of an inch wide. The strip rests on the

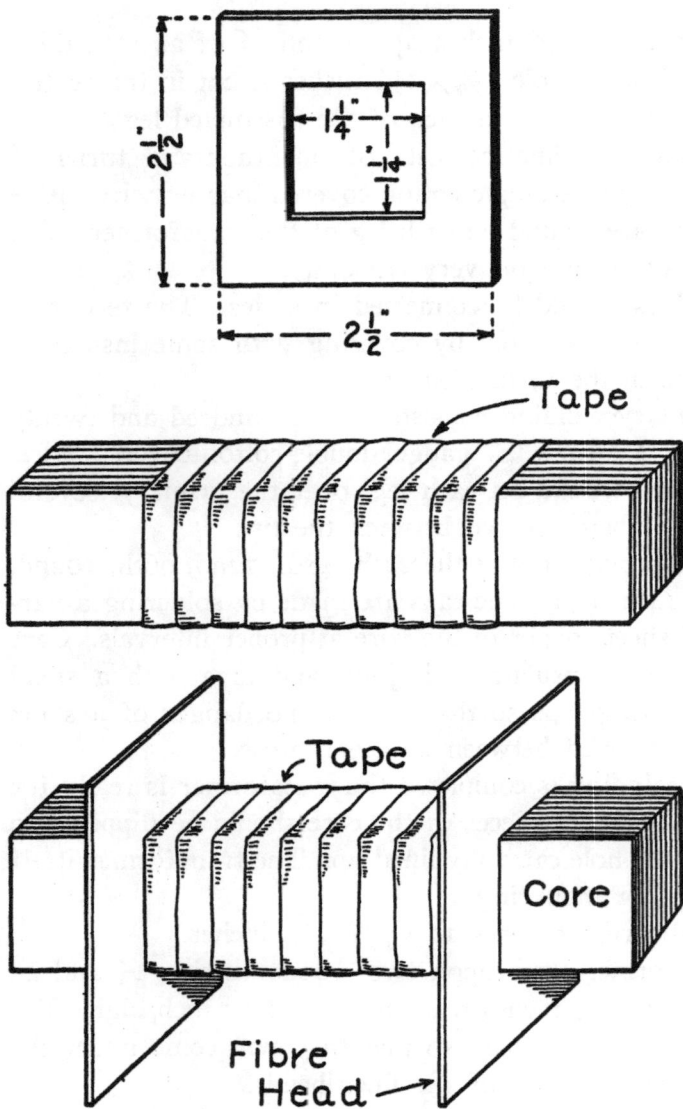

FIG. 177. — The Fiber Head is Shown at Top. In the Center is a Leg of the Transformer and at the Bottom, a Leg with Heads in Position for Winding.

ends of the core. The tie-rods are fastened on the under side of the base by means of a nut and washer on the ends. When the nuts are screwed up tightly, the cross-piece will pull the transformer firmly down to the base.

The regulating switches, two in number, are mounted on the lower part of the base. The contact points and the arm are cut out of sheet-brass, one-eighth of an inch thick. It is unnecessary to go into the details of their construction, because the dimensions are plainly shown in the illustration.

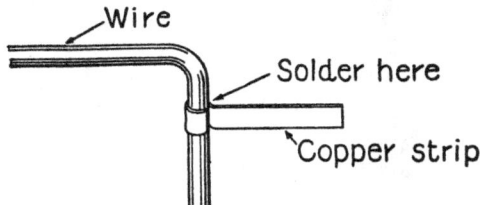

FIG. 178. — The Primary Taps Are Made by Soldering a Copper Strip to the Wire.

The contacts are drilled and countersunk so that they may be fastened to the base with small flat-headed wood screws.

Each switch-arm is fitted with a small knob to serve as a handle. The arm works on a small piece of brass of exactly the same thickness as the switch-points. Care must be taken that the points and this washer are all exactly in line, so that the arm will make good contact with each point. There are five points to each switch.

The switch, D, is arranged so that each step cuts in or out twenty-four turns of the secondary, the first point being connected with the end of the winding. The second point connects with the first tap, the third contact with the second tap, the fourth contact with the third tap, and the fifth contact with the fourth tap.

The switch, *E*, is arranged so that each step cuts in or out six turns. The contacts on this switch are numbered in the reverse direction. The fifth contact of switch *D*, and the fifth contact of switch *E*, are connected together. The fourth contact is connected to the fifth tap, the third contact to the sixth

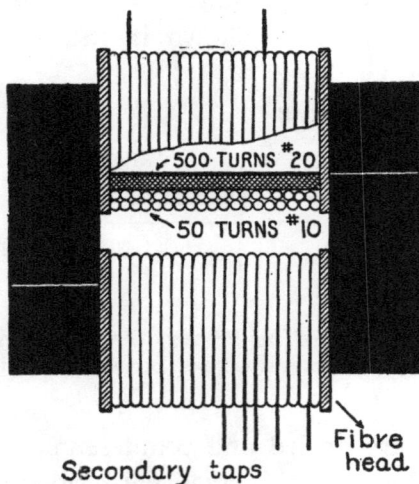

Fig. 179. — The Transformer, Completely Wound and Ready for Mounting.

tap, the second contact to the seventh, and the first contact to the end of the winding.

This arrangement makes it possible to secure any voltage from approximately one-half to ten in one-half-volt steps from the secondary of the machine. Each step on the switch, *D*, will give two volts, while those on *E* will each give one-half of a volt.

Two binding-posts (marked *P* and *P* in the drawing) mounted in the upper corners of the base are connected to the terminals of the primary winding. The two posts in the lower corners (marked *S* and *S* in the drawing) are connected to the switch levers, and are the posts from which the secondary or low voltage is obtained.

The transformer may be connected to the 110-V. alternating current circuit by means of an attachment plug and cord.

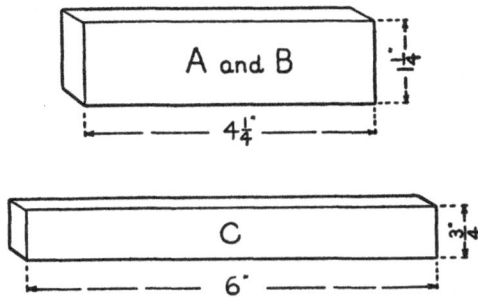

FIG. 180. — Wooden Strips for Mounting the Transformer on the Base.

One end of the cord is placed in each of the primary binding-posts. The other end of the cord is connected to the attachment plug so that the latter may be screwed into any convenient electric light socket.

The transformer must not be connected directly to the line. An instrument such as this is not designed for continuous service and is intended to be disconnected as soon as you are through using it.

It will be found a great convenience in operating many of the electrical devices described, wherever direct current is not essential.

An Electrolytic Rectifier

The electrolytic rectifier is a device for changing alternating current into direct current. It is fairly efficient if used only

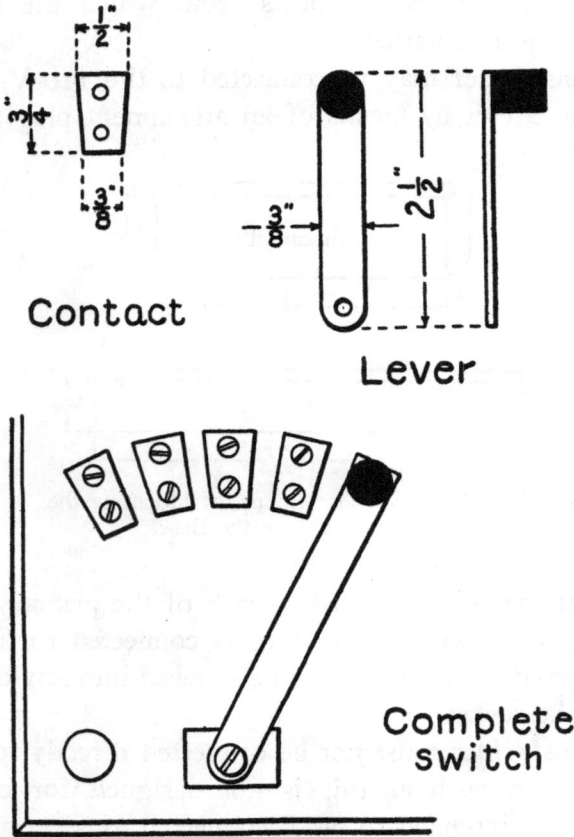

Contact

Lever

Complete switch

Fig. 181. — The Switch and Details of its Parts.

to rectify small amounts of current. It is not efficient and quickly becomes overheated when too much current passes through it.

A simple electrolytic rectifier consists of two electrodes, one of lead and the other aluminum, immersed in a suitable electrolyte.

A glass battery jar measuring 5×7 inches is preferable to a smaller jar because it will hold more electrolyte and not become hot as quickly.

A wooden cover supports the electrodes and also prevents evaporation. The cover may simply rest on the top of the

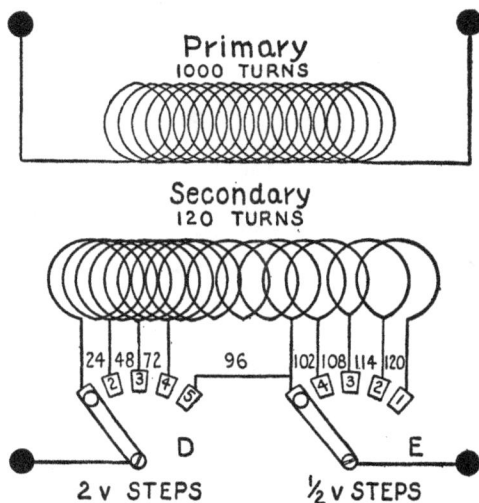

FIG. 182.—Diagram of Connections.

jar or have a groove turned in the underside so that it fits the rim of the jar snugly. Thoroughly saturate the wooden cover with paraffin by immersing it in a molten bath of that material. If the cover is allowed to remain in the bath until all bubbles have ceased to rise, the paraffin will thoroughly permeate the wood and protect it from the action of the chemical solution used in the rectifier.

The electrodes are cut out of thick sheet metal and made in

the form of strips about one and one-half inches wide and six inches long. The top of each electrode is bent over at right angles and bored with a $\frac{5}{32}$ inch hole so that an 8–32 brass machine screw will pass through. The screw serves not only to fasten the electrode securely to the underside of the cover but also as part of a binding post.

The electrolyte is a strong solution of sodium-bicarbonate.

The action of an electrolytic rectifier, in changing alternating current into direct current, is interesting. The rectifier

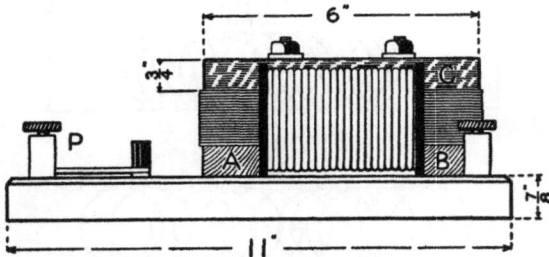

FIG. 183. — Side View of the Completed Transformer.

acts like an electrical valve which opens one way and closes the other.

If a battery is connected to the electrodes of a rectifier, the positive pole of the battery being connected to the lead and the negative to the aluminum, the current from the battery will flow through the rectifier and nothing unusual will happen.

If, however, the poles are reversed so that the positive pole is connected to the aluminum, oxygen gas will cover the surface and form a coating of aluminum oxide all over the electrode. Aluminum oxide is a poor conductor of electricity and it therefore forms an insulating coating which shields the electrode from the solution and stops the passage of current. This action is almost instantaneous. If the rectifier is connected to

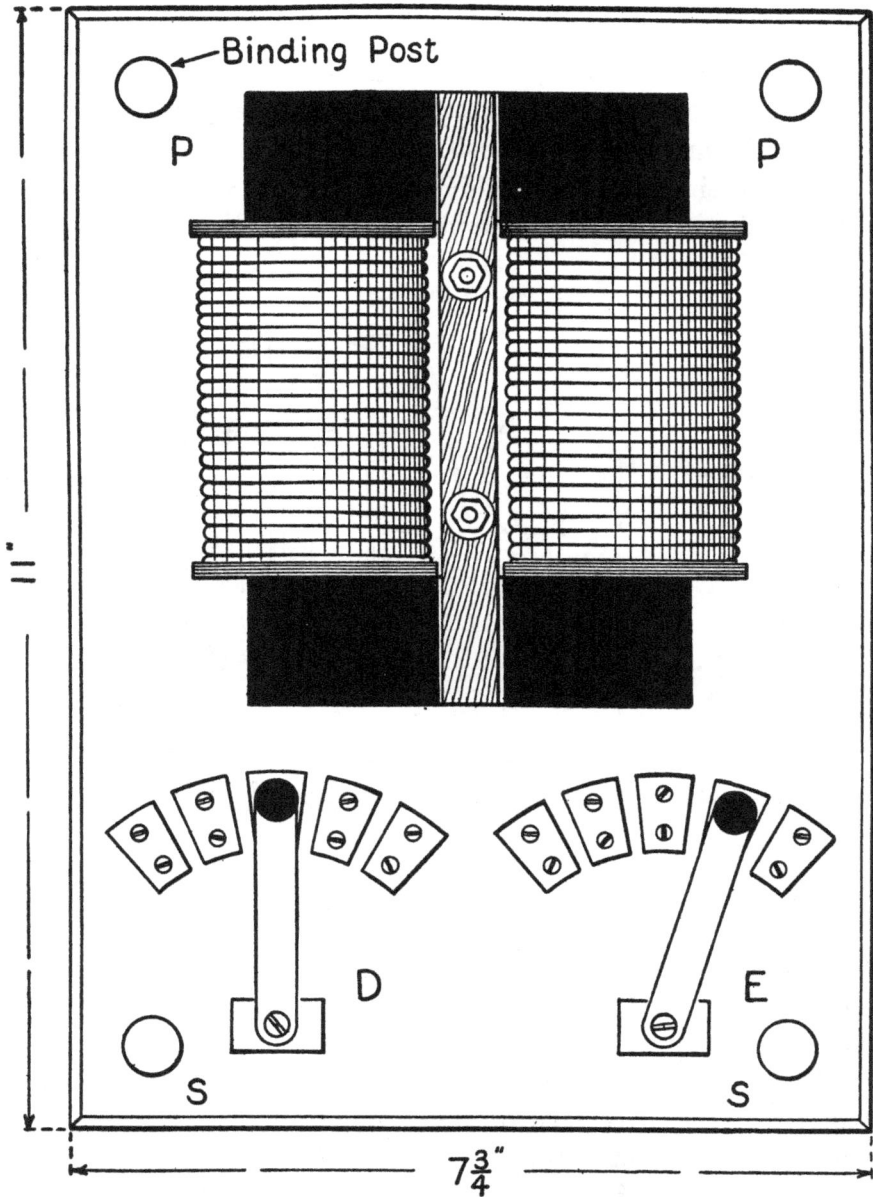

FIG. 184. — Top View of the Transformer.

an alternating current supply, it will act like a valve permitting the current to flow in one direction but stopping it whenever the aluminum electrode is positive.

When a single cell of rectifier is used on the 110 volt current supply, it should be connected in series with a lamp bank

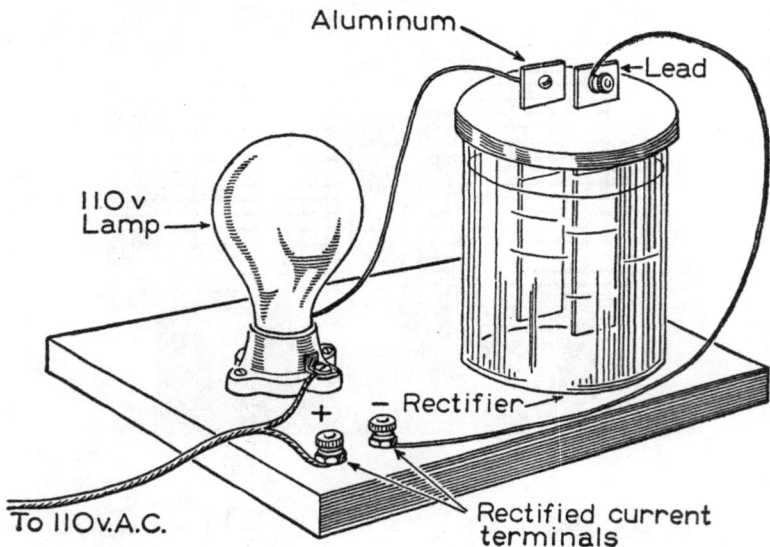

FIG. 185. — The Rectifier should be connected in series with a Lamp or a Lamp Bank. The most efficient operation is obtained by reducing the 110 volts with a Step-Down Transformer before it is rectified.

or a step-down transformer. The step-down transformer is more efficient and does not waste as much current as a lamp bank.

When using an electrolytic rectifier to recharge a storage battery the negative pole of the storage cell must always be connected to the aluminum electrode of the rectifier.

A single cell of rectifier connected in series with a lamp bank or a transformer only utilizes one-half of the alternating current cycle. It is possible by means of four cells, to utilize the complete cycle.

An electrolytic rectifier composed of cells of the size just described will not efficiently handle a current of more than

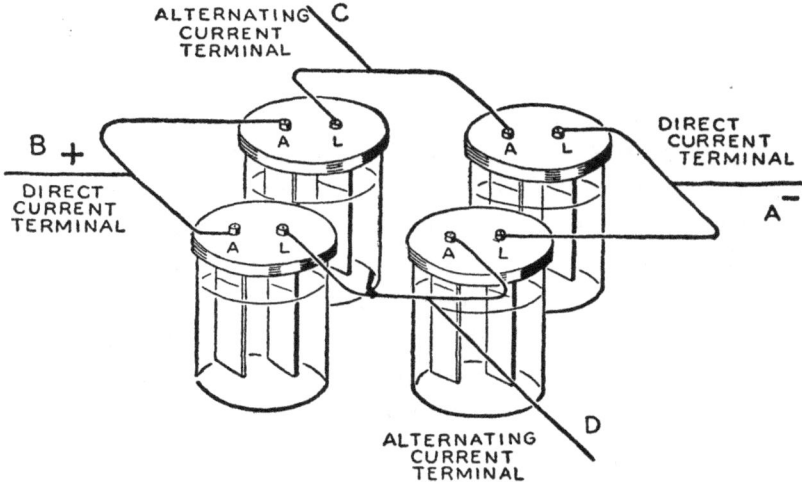

FIG. 186. — Four Cells of Electrolytic Rectifier Connected to Utilize the Complete Alternating Current Cycle.

two amperes. When the solution in the rectifier becomes very hot, it will not operate as efficiently as when cold. The solution becomes exhausted after a certain amount of use and requires renewal.

CHAPTER XIV

WIRELESS TELEGRAPHY

No branch of electrical science ever appealed more to the imagination of the experimenter than that coming under the heading of *wireless* telegraphy. Twenty-five years ago when the first edition of this book was published, there were to be seen the ear-marks of *amateur* wireless *telegraph* stations in the aerials and masts set up in trees and on housetops. It is estimated that there were nearly a quarter of a million such stations in the United States. They were not for the purpose of listening to the entertainment furnished by radio broadcast programs, for at that time radio telephony was not practical, and broadcasting of regular programs had not been thought of, but were "wireless" stations for sending and receiving wireless *telegraph* signals.

These amateur stations could listen to the telegraph messages passing between ships at sea and the shore or communicate with one another over distances varying from two or three miles to halfway across the continent. Those stations which were equipped for transmitting were licensed by the United States Department of Commerce. No other nation had liberal laws which encouraged the operation of private stations. Consequently during the World War, the wireless telegraph equipment and operators of the American Army and Navy were vastly superior to those of the other warring nations solely because of the stimulus and experience of the large number of amateur operators that had been developed in this country.

The amateur radio operator was also undoubtedly directly responsible for modern broadcasting, because amateurs were the first to send out phonograph music more or less regularly over their experimental wireless telephone sets, thus creating

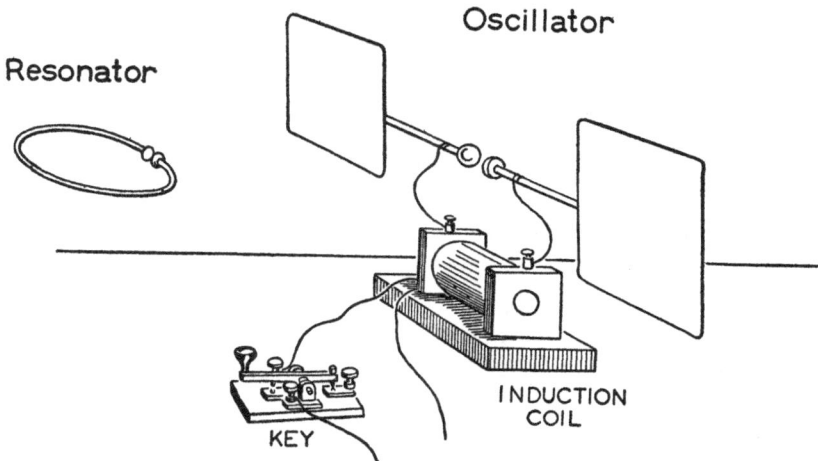

THE FIRST MAN-MADE ELECTRIC WAVES

FIG. 187. — Heinrich Hertz was first to discover how to create electro-magnetic waves. He produced electric sparks with an induction-coil. The sparks took place between two metal balls fitted with plates or vanes. The sparks generated electric waves which could be detected by a "resonator." The resonator was simply a metal ring which was not quite completely closed. When the ring was held near the induction-coil it was struck by electromagnetic waves and minute sparks jumped across the almost-meeting ends of the ring. *From "Getting Acquainted with Radio" by Alfred Morgan. Courtesy D. Appleton-Century Co.*

the idea of broadcasting. They constituted a vast army of listeners already equipped with sets to receive the first broadcast programs when they were finally started on a commercial basis.

In the days before broadcasting, the term *radio* was not universally used. The word "wireless" was applied instead. I am still going to use the older word when referring to wireless or radio-*telegraphy,* and *radio* only when the more modern transmission and reception of speech or music by means of the three-element vacuum tube with which we are familiar in the broadcast receiver, is involved.

The really valuable and most practical use for wireless telegraphy was the transmission of messages between ships at sea or between ships and the shore. Prior to the advent of wireless, ships at sea were practically isolated from the rest of the world unless they were within sight of land. The value of wireless in making the sea safer for travel by furnishing a means by which storm warnings, time signals and distress calls could be sent, to say nothing of the convenience to passengers and steamship companies in the transaction of their personal affairs and business, is inestimable.

No development of electrical science is as interesting or involves as many principles of electricity and physics as *radio.* *Wireless telegraphy was the forerunner of modern radio.*

There is really no great mystery about this wonderful art which made possible the instantaneous transmission of messages over immense distances without any apparent physical connection save that of the earth, air, or water.

Did you ever throw a stone in a pool of water? As soon as the stone struck, little waves spread out from the spot in gradually enlarging circles until they reached the shore or died away.

By throwing several stones in succession, with varying intervals of time between them, it would be possible so to arrange a set of signals, that they would convey a meaning to a second person standing on the opposite shore of the pool.

Wireless telegraphy is based upon the principle of *creating and detecting electrical waves.*

In order to explain electrical waves scientists have suggested

that all space is filled with an "imaginary" substance called
ether. The ether is invisible, odorless, and practically weightless. This ether, however, bears no relation to the anaesthetic
of that name which is used in surgical operations.

It is supposed to surround and penetrate all substances and
all space. According to theory it exists in a vacuum and in solid
rocks. Since the ether does not make itself apparent to any of

FIG. 188. — Little Waves Spread Out From the Spot. Electric
Waves May be Likened to Ripples in a Pond.

our physical senses, some of these statements may seem con-
tradictory. Its definite existence cannot be proved except by
reasoning, but by accepting and imagining its reality, it is possible to understand and explain many scientific puzzles.

A good instance is offered by the sun. Light and heat can
be shown to consist of extremely rapid vibrations. That fact
can be proved. The sun is over 90,000,000 miles away from
our earth and yet light and heat come streaming down to us
through a space that is devoid even of air. Something must
exist as a medium to transmit these vibrations; it is the ether.

Let us consider again the pool of water. The waves or

ripples caused by throwing in the stone are vibrations of the water. The distance between two adjacent ripples is called the *wave length*.

The distances between two vibrations of light can also be *measured*. They are so small, however, that they may be spoken of only in *thousandths* of an inch. The waves created in the ether by wireless telegraph apparatus are the same as those of light except that their length usually varies from 75 to 9,000 *feet* instead of a fraction of a thousandth of an inch.

The first wireless telegraph transmitters consisted of a telegraph key connected in series with a set of cells and the *primary*

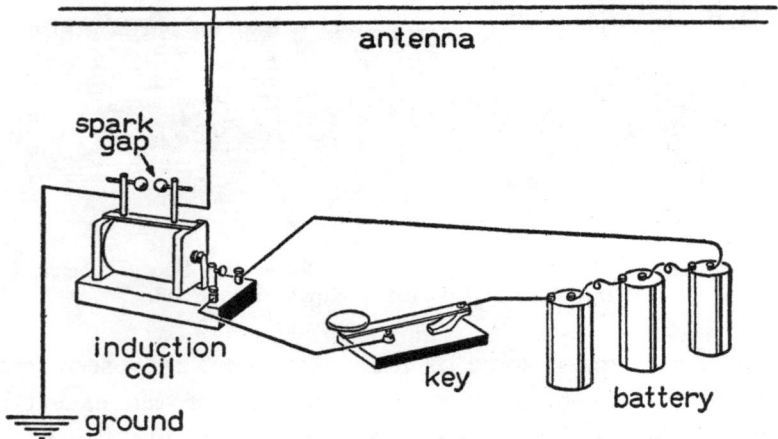

FIG. 189. — The first Radio Transmitter was called a Wireless Telegraph and consisted of an Induction Coil connected to an Aerial and Ground.

of an induction coil, which, it will be remembered, is simply a coil consisting of a few turns of wire. This induces a high voltage in a second coil consisting of a larger number of turns and called the *secondary*.

The terminals of the secondary wire led to a spark-gap—

an arrangement composed of two polished brass balls, separated by a small air-gap. One of the balls, in turn, was connected to a metal plate buried in the earth, and the other to wires suspended high in the air and insulated from all surrounding objects.

Here is an explanation of what happens. When the key at the transmitter is pressed, the battery current flows through the primary of the induction coil and generates in the secondary a current of very high voltage, 20,000 volts or more, which is able to jump an air-gap in the shape of a spark at the secondary terminals. The latter are connected to the earth and aerial, as explained above. The high potential currents are therefore enabled to charge the aerial. The charge in the aerial exerts a great tendency to pass into the ground, but is prevented from doing so by the small air-gap between the spark-balls until the charge becomes so great that the air-gap is punctured and the charge passes across and flows down into the ground. The passage of the charge is made evident by the spark between the two spark-balls.

The electrical charges flowing up and down the aerial disturb the ether, strike it a blow, as it were. The effect of the blow is to cause the ether to vibrate and to send out waves in all directions. It may be likened to the pond of water which is suddenly struck a blow by throwing a stone into it, so that ripples are immediately sent out in widening circles.

These Waves in the Ether are called *electro-magnetic* or *Hertzian* waves, after their discoverer, Hertz. The distance over which they pass is dependent upon the power of the transmitting station. The waves can be made to correspond to the dots and dashes of the telegraphic code by so pressing the key. If some means of detecting the waves is employed we may readily see how it is possible to send wireless messages.

The Action of the Receiving Station is just the opposite of that of the transmitter. When the waves pass out through the ether, some of them strike the aerial of the receiving station and generate a charge of electricity in it which tends to pass down into the earth. If the transmitting and receiving stations are very close together and the former is very powerful, it is possible to make a very small gap in the receiving aerial across

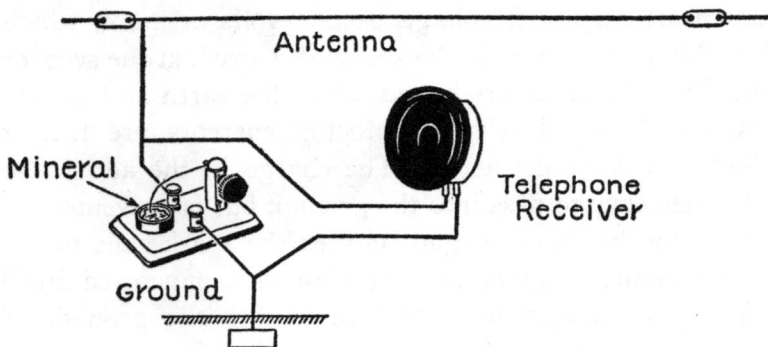

Fig. 190. — A Simple Radio Receiver.

which the charge will jump in the shape of sparks. Thus the action of the receptor simply takes place in a reversed order from that of the transmitter.

If the stations are any considerable distance apart, it is impossible for the currents induced in the receiving aerial to produce sparks, and so some more sensitive means of detecting the waves from the transmitter is necessary, preferably one which makes itself evident to the sense of hearing.

The telephone receiver is an extremely sensitive instrument, and it requires only a very weak current to operate it and produce a sound. The currents or *oscillations* generated in the aerial, however, are *alternating* currents of *high frequency*, that is, they flow in one direction and then reverse and flow in the other several thousand times a second. Such a current cannot be made to pass through a telephone receiver, and in order to do so the nature of the current must be changed by converting it into *direct* current flowing in one direction only. This is called *rectifying* the current.

Certain crystalline minerals possess the remarkable ability to change or rectify an alternating current into a direct current. They act as sort of "an electrical valve," allowing the current to flow in one direction but not in the other. One half of the alternating current cycle only can therefore pass through one of these mineral rectifiers or detectors.

The *crystal detector* consists of a sensitive mineral placed between two contacts and connected so that the aerial currents must pass through it on their way to the ground. A telephone receiver is connected to the detector so that the *rectified* currents (currents which have been changed into direct current) pass into it and produce a sound. By varying the periods during which the key is pressed at the transmitting station, according to a prearranged code, the sounds in the receiver may be made to assume an intelligible meaning.

How to Build a Crystal Receiving Set

In order simply to "listen in" and receive the messages of other near-by amateurs and ships or commercial stations it is only necessary to employ:

> A Detector
> A Tuning Coil
> A Fixed Condenser
> And a Telephone Receiver

connected to a suitable aerial and ground.

Such an outfit is described in the following pages. It may be used to pick up radio programs from nearby broadcasting stations.

One of the first things to receive attention when thinking of building a wireless or radio outfit should be

The Antenna

Every wireless station is provided with a system of wires elevated high in the air, above all surrounding objects, the purpose of which is to radiate or intercept the electromagnetic

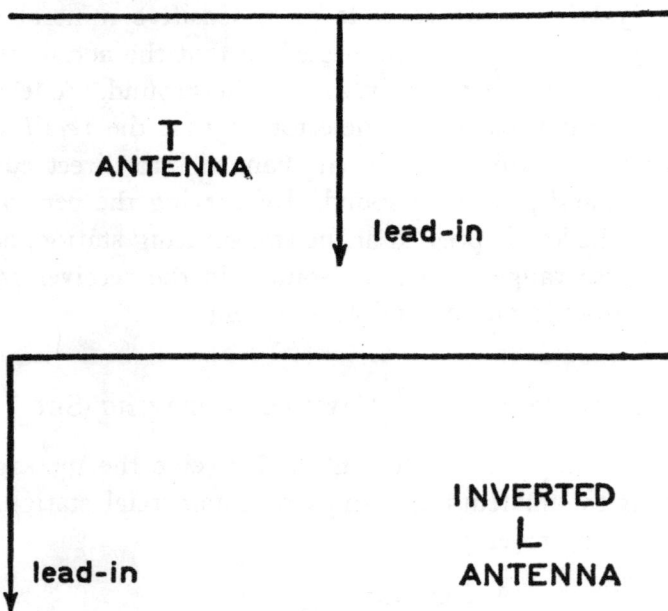

T
ANTENNA

lead-in

INVERTED
L
ANTENNA

lead-in

Fig. 191. — Two common Types of Receiving Antennas.

waves, accordingly as the station is transmitting or receiving. This system of wires is, as already has been stated, called the *aerial* or *antenna*.

The arrangement of the antenna will greatly determine the efficiency and range of the apparatus.

The antenna should be as long as it is reasonably possible to make it, that is from 50 to 100 feet.

It will be necessary for most amateurs to put up their antenna in some one certain place, regardless of what else may be in the vicinity, but whenever possible the site selected should preferably be such that the antenna will not be in the immediate neighborhood of any tall objects, such as trees, smoke-stacks, telephone wires, etc., because such objects will interfere with the antenna and noticeably decrease the range of the station.

Bare copper wire or bronze cable makes the best aerials. Iron wire should never be used for an aerial, even if galvanized or tinned, because it tends to choke the currents which must flow up and down the aerial when the station is in operation.

The antenna must be very carefully insulated from its supports and all surrounding objects. The insulation must be strong enough to hold the weight of the antenna and able to withstand any strain caused by storms.

Special antenna insulators made of molded glass or porcelain and having an eye in each end are the best.

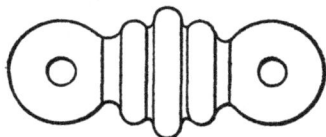

Fig. 192. — Antenna Insulator.

Inexpensive antenna insulators made of porcelain or glass are to be found in almost any 5 and 10 cent store or radio shop.

Ordinary porcelain cleats may be used on small antennas where the strain is light.

One insulator should be placed at each end of each wire.

Transmitting antennas are made up of several wires, but the receiving aerial need consist of one wire only.

Most antennas are supported from a pole placed on the top of the house, in a tree, or erected in the yard.

Much of the detail of actually putting up an aerial or antenna must be omitted, because each experimenter will usually meet different conditions.

It should be remembered, however, that the success of the whole undertaking will rest largely upon the construction of a proper antenna. The most excellent instruments will not give

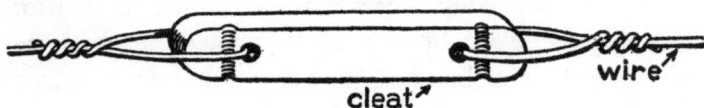

Fig. 193. — A Porcelain Cleat Will Make a Good Insulator for Small Aerials.

very good results if connected to a poor antenna, while, on the other hand, inferior instruments will often give fair results when connected to a good antenna.

The antenna should be at least thirty feet high.

The wire should not be smaller than No. 14 B. & S.

The masts which support the antenna should be of wood and provided with pulleys so that the wires may be lowered any time it may be necessary. The mast should be thoroughly braced with stays or guys so as to counteract the strain of the antenna.

The antenna should not be hoisted up perfectly tight, but should be allowed to hang somewhat loose, as it will then put less strain on the ropes and poles that support it.

When an antenna is to be fastened in a tree, it is best to attach it to a pole placed in the top of the tree, so that it will come well above any possible interference from the branches.

The wire leading from the antenna to the instruments should be very carefully insulated throughout its length. This part of the antenna is called the "lead-in."

It is very important that a good ground connection be secured for wireless instruments, for it is absolutely necessary for the proper working of the apparatus. Amateur experi-

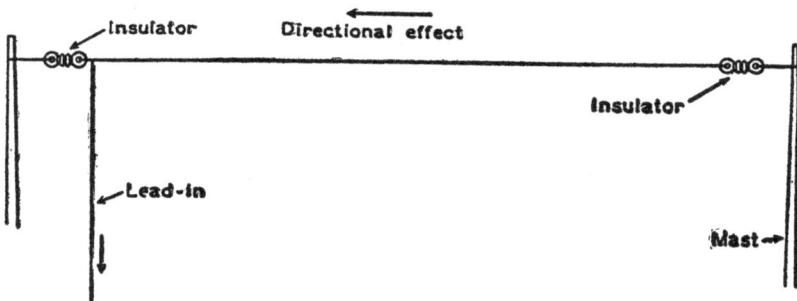

FIG. 194. — Complete Antenna of the Inverted L Type.

menters usually use the water or gas pipes for a ground, and fasten the wires by means of a ground clamp. In the country, where such pipes are not available, it is necessary to bury a sheet of copper, three or four feet square, in a moist spot in the earth and connect a wire to it.

When the aerial and ground connections have been arranged you will be ready to commence the construction of the

Tuning Coil, which is a very simple arrangement making it possible to receive messages from greater distances, and also to somewhat eliminate any messages not desirable and to listen without confusion to the one wanted.

A tuning coil consists of a single layer of wire wound upon a cylinder and arranged so that connection may be had with any part of it by means of sliding contacts.

The cylinder upon which the wire is wound is a cardboard

tube six and three-quarters inches long and two and seven-eighths inches in diameter outside. It should be given two or three coats of shellac both inside and out so that it is thoroughly impregnated, and then laid away until dry. This treatment will prevent the wire from becoming loose after the tube is wound, due to shrinkage of the cardboard.

After having become dry, the tube is wound with a single layer of No. 25 B. & S. gauge green silk or cotton-covered mag-

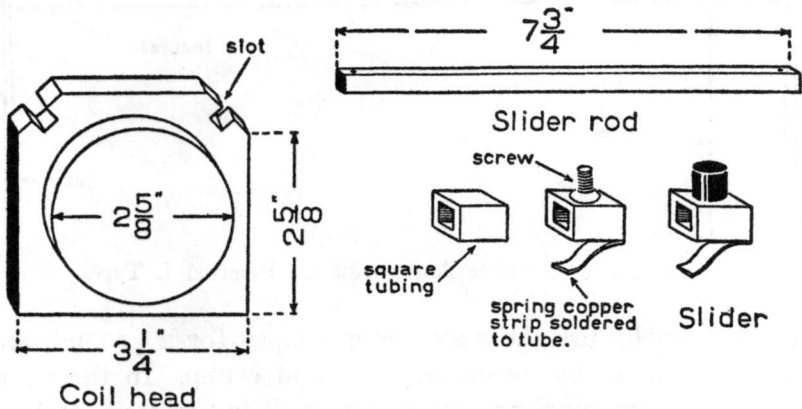

slot

$7\frac{3}{4}$"

Slider rod

screw

square tubing

spring copper strip soldered to tube.

Slider

$2\frac{5}{8}$"

$2\frac{5}{8}$"

$3\frac{1}{4}$"

Coil head

Fig. 195. — Parts of the Tuning Coil.

net wire. The wire must be wound on very smoothly and tightly, stopping and starting one-quarter of an inch back from each end. The ends of the wire are fastened by weaving back and forth through two small holes punched in the cardboard tube with a pin.

The winding should be given a single coat of clear varnish or white shellac and allowed to dry.

The coil heads or end pieces are cut from one-half-inch wood according to the plan and dimensions shown in the accompanying illustration.

The top corners are beveled and notched to receive the slider-rods. A circular piece of wood two and five-eighths inches in diameter and three-eighths of an inch thick is nailed to the inside of each of the coil heads to support the ends of the cylinder.

The wooden parts should be stained mahogany or some other dark color and finished with a coat of shellac or varnish.

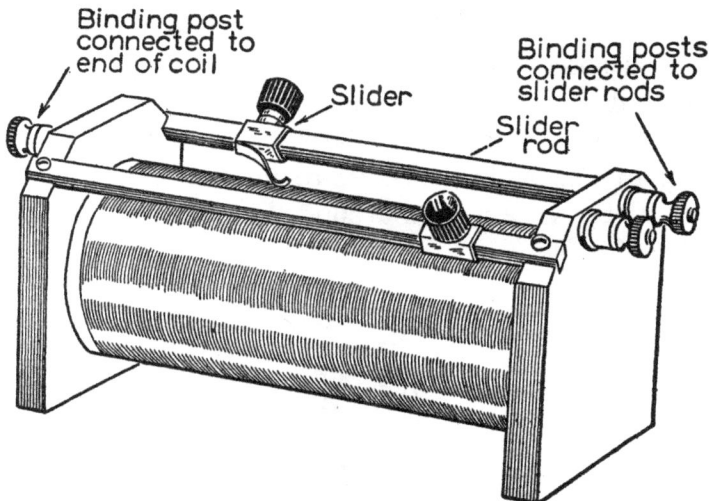

FIG. 196. — Complete Double-Slider Tuning Coil.

The slider-rods are square brass $\frac{3}{16} \times \frac{3}{16}$ inches and seven and three-quarters inches long. A small hole is bored near the ends of each, one-quarter of an inch from the edge, to receive a round-headed brass wood screw which holds the rod to the tuner end.

The slider is made from a small piece of brass tubing, three-sixteenths of an inch square. An 8–32 flat-headed brass screw is soldered to one face, in the center. A small strip of phosphor bronze sheet or spring copper soldered to the bottom of the slider forms a contact for making connection to the wire on the cylinder. A small knob screwed to the slider makes a neat and efficient handle.

Two sliders are required, one for each rod.

The cardboard tube is held in place by several small brass nails driven through it into the circular pieces on the coil heads.

A slider is placed on each of the slider-rods and the rods fastened in the slots in the coil ends by a small round-headed brass screw, passing through the holes bored near the ends for that purpose.

Two binding-posts are mounted on one of the coil ends. One should be connected to each of the slider-rods. A third binding-post is placed below in the center of the head and connected to one end of the wire wound around the cylinder.

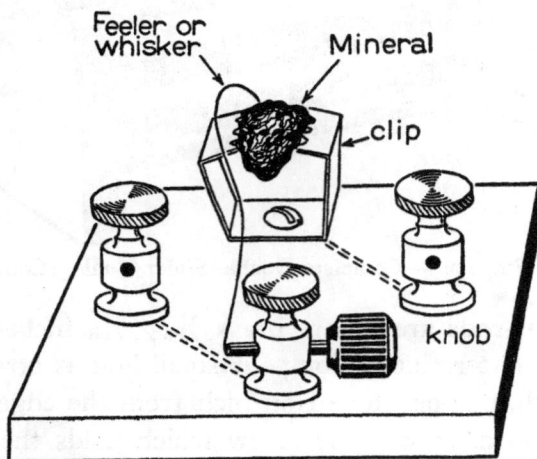

FIG. 197. — "Cat-Whisker" Detector.

A small, narrow path along the coil, directly underneath each slider and to which the copper strip can make contact, must be formed by scraping the insulation off the wire with a sharp knife. The sliders should make contact with each one of the wires as they pass over, and should slide smoothly without damaging or disarranging any of the wires.

When scraping the insulation, be very careful not to loosen the wires or remove the insulation from between them, so that they are liable to short-circut between adjacent turns.

Detectors are very simple devices and consist merely of an arrangement for holding a small piece of certain minerals and making a contact against the surface.

The detector often termed a "cat-whisker," because of the long, fine wire resting on the mineral is one of the most sensitive.

A simple form consists of a small clip, formed by bending a strip of sheet-brass, which grips a piece of galena.

Galena may be obtained from any dealer in radio supplies. A piece of No. 30 phosphor bronze wire is soldered to the end

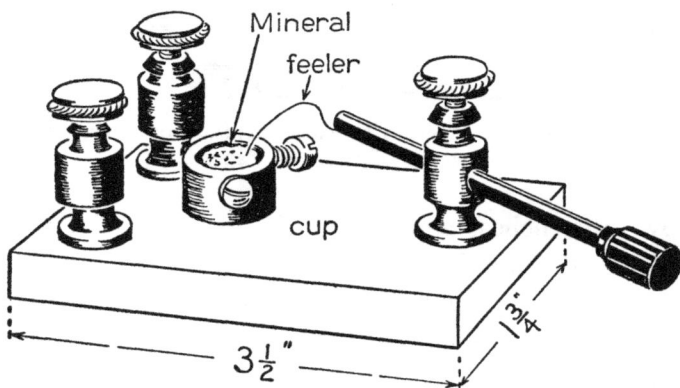

FIG. 198. — Another Form of the "Cat-Whisker" Detector.

of a short length of brass rod supported by a binding post. This part of the detector is called the "feeler" or "whisker." The other end of the rod is fitted with a knob.

The detector is fitted with binding posts and may be mounted upon any suitable small base. The mineral clip is connected to one post and the binding-post supporting the "feeler" to the other. The tension or pressure of the end of the fine wire upon the mineral may be regulated by twisting the electrose knob so as to twist the rod. The different portions of the crystal may be "searched" for the most sensitive spot by sliding the rod back and forth.

Fixed Condenser. The construction of the condenser is illustrated. Take twenty-four sheets of thin typewriter paper,

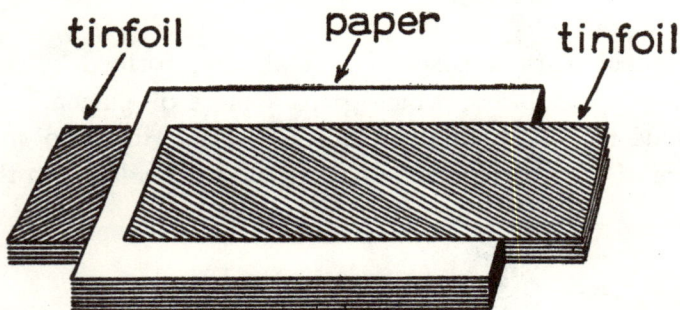

Fig. 199. — The Fixed Condenser consists of alternate Sheets of Tinfoil and Paper.

three by four inches, and twenty-three sheets of tinfoil, two by four inches. Pile them up, using first a sheet of paper then a sheet of tinfoil then paper, and so on, so that every two sheets of tinfoil are separated by a sheet of paper. Each sheet of tinfoil must, however, project out beyond the edge of the paper. Connect all the tinfoil projections on one end of the condenser together and attach a small wire. Connect all those on the opposite side in a similar manner. Put a piece of cardboard on the top and bottom. Then fasten a couple of rubber bands around the condenser to hold it together.

If it is desired to give the condenser a finished appearance, it may be placed in a small box provided with binding-posts to which the terminals of the condenser are connected. The box may be wood or cardboard. Keep the condenser flat and tightly pressed together.

Telephone Receivers for use with wireless instruments must be purchased. Their construction is such that they cannot be made by the experimenter.

FIG. 200. — A Radio Headset.

A seventy-five ohm, double-pole telephone receiver will do for stations not wishing to receive farther than fifty miles.

In order to secure the best results from wireless instruments, it is necessary to have receivers especially made for wireless. Each receiver should have 1000 ohms resistance. Two receivers, a double headband, and a double cord form a complete head set.

Connecting the Receiving Apparatus

Many experimenters may wish to build a wireless receiving set which is permanently connected and in which the instruments are so mounted that they are readily portable and may

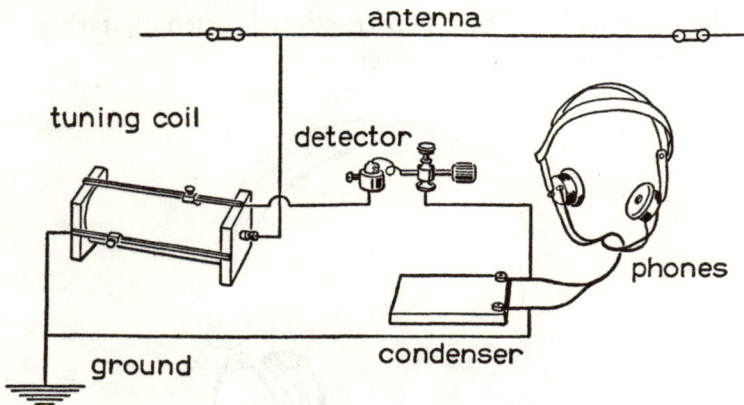

Fig. 201. — A Circuit Showing How to Connect a Double-Slider Tuning Coil.

be easily shifted from one place to another without having to disturb a number of wires.

The base is of wood, and is nine inches long, seven inches wide, and one-half of an inch thick.

A double-slider tuning coil is fastened to the back part of the base by two small wood-screws passing upwards through the base into the tuner heads.

The fixed condenser is enclosed in a rectangular wooden block which is hollowed out underneath to receive it and then screwed down to the base in the forward right-hand corner.

The crystal detector is mounted in the forward left-hand part of the base.

The telephone receivers are connected to two binding-posts mounted alongside the detector.

The wires which connect the various instruments should be

passed through holes and in grooves along the under side of the base so that they are concealed.

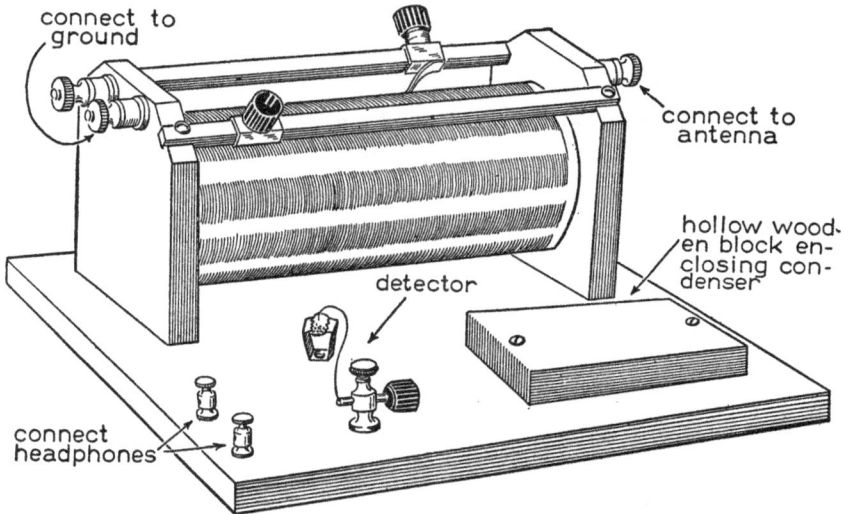

FIG. 202. — A Receiving Set consisting of Tuner and Crystal Detector.

THE CONTINENTAL CODE

The Continental Code is the one usually employed in wireless telegraphy. It differs slightly from Morse as it contains

FIG. 203. — The Continental Alphabet used in Sending Radio Telegraph Messages.

no space letters. It will be found easy to learn and somewhat easier to handle than Morse.

Two or three months' steady practice with a friend should enable the young experimenter to become a very fair wireless telegraph operator. Then by listening for some of the commercial or ship wireless stations during the evening it should be possible to become very proficient.

CHAPTER XV

RADIO RECEIVING SETS

THE building and adjustment of radio equipment for receiving broadcast programs is a subject large enough to require more than a whole volume. There are a great many different basic circuits and combinations of circuits. You have all probably heard of *neutrodynes, super-heterodynes, radio-frequency amplification, regenerative circuits,* etc. Each one of these might even be worthy of a volume of its own. Therefore, I hope that when you turn to this chapter, you will not be disappointed if you do not find a detailed explanation of many of the radio terms and devices which you have seen advertised or heard about. If you wish to build an *elaborate* radio receiver which you may consider to be the *latest* in design, I would advise that you follow the instructions given in some of the radio magazines or purchase one of the booklets devoted to the details of building a certain type of receiver. There are styles and fads in radio just as in other things. These are to a large extent created by the manufacturers of parts and the publishers of radio magazines and columns in order to increase business. There are improvements, of course, from time to time in radio, but ninety-nine times out of a hundred the so-called new circuits are basically not new, just disguised and modified forms of old ideas. Only a book which is reprinted several times a year could keep up with the "latest improvements."

However, the three-electrode vacuum tube is universally

used in modern radio and is the basis of the present-day art of transmitting and receiving sounds. Incidentally, long distance wire telephony also depends upon it.

If you wish to understand radio, you must first understand the three-electrode vacuum tube. The elementary principles of its action and operation are ignored in most articles dealing with radio and the limited space which can be given to radio in this book can therefore perhaps be made of inestimable value to you by devoting it largely to an explanation of the interesting "happenings inside of the electric lamp" which sends and receives sounds over long distances, transmits pictures through the air, and is the basis of television or seeing by wire and radio.

It is hardly possible to explain the action of the vacuum tube in really simple language. You will have to study this portion of "The Boy Electrician" a little more carefully than the rest. But if you do, and thereby gain an understanding of this wonderful little device, you will be able to build and operate a radio set far better and more intelligently than heretofore.

A receiving set in the early days of this still young art of radio was and still is (as I have explained in the last chapter) an arrangement of the following essential parts:

1. An aerial
2. A tuning device
3. A detector and amplifier
4. Head-telephones or receivers or loud speaker
5. A ground connection
6. A suitable current supply

Sounds, as of the voice, or a musical instrument, are transmitted by *radio* for great distances by the medium of electromagnetic waves created by electrical vibrations. How this is accomplished in wireless telegraphy has already been explained. The principles used in radio are the same.

Originating at a sending or broadcasting station, these waves carrying the music or speech also radiate in every direction at a rate of vibration or at a *frequency* measured by the number of waves which pass a given point in a second, amounting sometimes to several million and travelling with the speed of light, or 186,000 miles a second.

When these waves occur more rapidly than 20,000 per second, they are arbitrarily said to be of *radio* frequency, and cannot be heard by the human ear. When waves of radio frequency (above 20,000 per second) strike the aerial of a distant receiving set, they impose upon that aerial, waves of the same high frequency, and it is the function of the receiving apparatus to convert them into waves of lower frequency, or below 20,000 per second *so that they may be heard by the human ear,* or, in other words, become *audible.* When thus converted, they are called *audio*-frequency waves.

Waves of radio-frequency are sent out by an alternating current, and in turn produce alternating currents. Before they will transmit sound, their frequency or rate of alternation must not only be reduced to the audible state, below 20,000 per second, but the current must be transformed from an alternating current to a pulsating direct current through which alone sound may be carried to the ear.

This is the function of the *detector,* or as it might also be called, the *rectifier.* It rectifies, technically speaking, or more plainly, changes the radio-frequency vibrations coming through the receiving aerial into vibrations of audio-frequency.

For many years it had been known that a piece of red hot metal will give off millions of minute particles or negative charges of electricity, called *electrons,* in much the same manner that a piece of iron heated white in a blacksmith's forge throws off sparks. When the red-hot metal is enclosed in a vacuum, the quantity of electrons given off is greatly increased. Several years before radio was an accomplishment or perhaps even thought of, Thomas A. Edison performed an interesting experiment and made a very valuable discovery. He placed a metallic plate inside an electric-light bulb and discovered that when the filament of the bulb was lighted or heated, an elec-

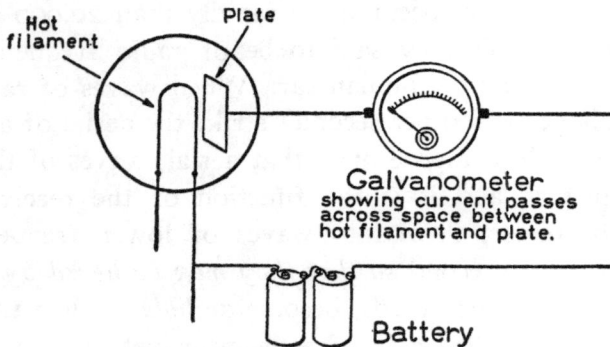

Fig. 204. — Edison's Experiment from which in later years the Radio Tube was developed.

tric current would flow across the space between the filament and the plate. When the filament was cold, no current would pass. The diagram in Figure 204 shows the arrangement of the apparatus in this famous experiment which was known as the discovery of the "Edison effect." No practical use of the "Edison effect" was made in radio until the English scientist, Fleming, took a vacuum tube with a heated filament like that of an electric light and placed a metallic plate in the zone of the myriads of electrons emitted by the filament and so discovered that by a sort of valve action, it would pass electricity better in one direction than in the opposite direction. By placing the valve in series with a source of radio-frequency oscillations, he found that as in the case of the mineral detector used in wireless telegraphy, one-half of each oscillation could be suppressed, and the circuit would then be traversed by a pulsating direct current, and the desired sounds produced in the receivers. This new *glow-lamp* or *oscillation valve,* as he called it, had the property of rectifying alternating current very much

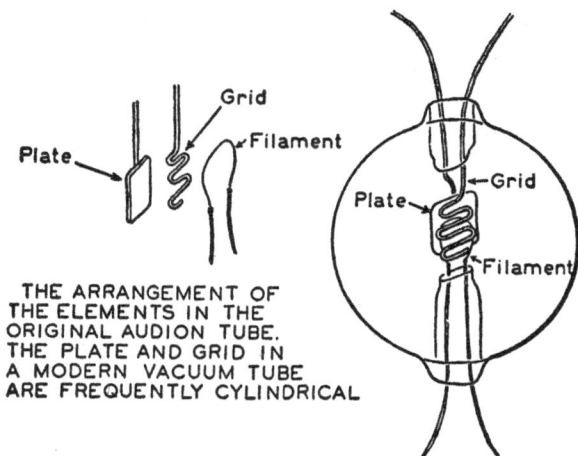

THE ARRANGEMENT OF THE ELEMENTS IN THE ORIGINAL AUDION TUBE. THE PLATE AND GRID IN A MODERN VACUUM TUBE ARE FREQUENTLY CYLINDRICAL

FIG. 205. — The DeForest Audion.

better than the crystal or mineral which had previously been used as a detector.

In 1907, Lee DeForest made an epoch-marking invention when attempting to improve Fleming's oscillation valve. He added a third element in the form of a metallic cage or spiral of wire called the *grid.* This important invention resembles an incandescent electric lamp in appearance, and, like the lamp, is highly exhausted of air or gas, but contains in addition to the metallic filament which is electrically heated to incandescence,

1. a cold metal plate placed apart from the filament and having an external connection,

2. a cold metallic cage or spiral wire called the grid, placed between the filament and plate and apart from both.

It was called the *audion* by DeForest, and its modern form (Radiotron) is now in universal use as the device which not only detects radio signals but also *generates* them as well.

The diagram in Figure 206 shows a simple audion receiving circuit such as was in use prior to the invention of the "regenerative" circuit described later.

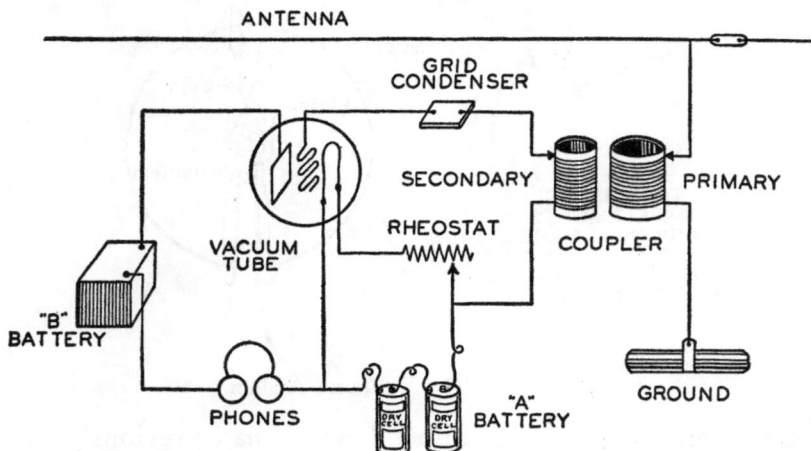

FIG. 206. — Simple Radio Receiver using a Vacuum Tube as the Detector.

The aerial is connected to one end of a coil of wire constituting the primary of a transformer or "coupler." The other end of the primary is connected to the ground. The oscillations generated in the aerial by incoming signals thus have to pass through the primary of the transformer, generating corresponding currents in the secondary. The secondary of the transformer is connected to the filament and grid (marked *F* and *G*) of the audion. The grid condenser is simply a very small condenser made of mica and tinfoil sheets. The filament is lighted by a six-volt storage battery called the "A" battery, whose current can be regulated by a small rheostat "R" and the brilliancy of the filament thereby controlled. The space between the filament and the plate is made part of a circuit (the plate circuit) which includes a pair of telephone receivers and a battery, called the "B" battery.

The alternating currents set up in the antenna by the incom-

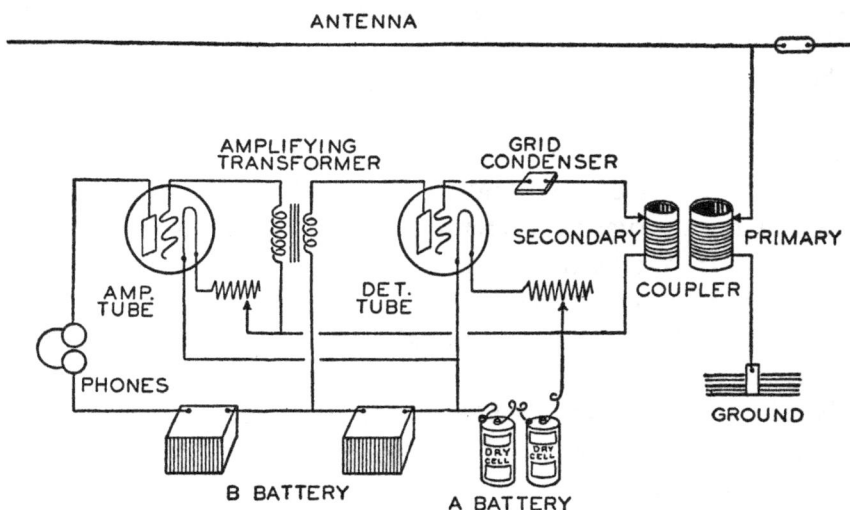

FIG. 207. — The Same Circuit Shown in Figure 206 but With a Second Audion Tube Added as an Amplifier.

ing signals are transferred to the grid by the "coupler" and cause the grid to become alternately positive and negative. The alternate positive and negative charges on the grid correspondingly decrease and increase the current from the "B" battery which can pass across the space between the hot filament and the plate, thus producing a pulsating direct current and setting up sounds in the telephone receivers.

The valve action of the grid was a great improvement over

all other forms of detectors but still had its limits in efficiency. The incoming oscillations, after travelling many miles from a broadcasting station, lose power as they proceed through the air and die down so that when they reach the receiving station they are often very feeble and do not set up fluctuations of the current in the telephone receivers which are strong enough to be useful. This led scientists to seek for some means of improving the amplifier, and although many men in different parts of the world made the same discoveries at about the same time, the courts, after much litigation of the patents in question, have decided that DeForest was also the inventor of the *audion amplifier*.

The amplifier circuit shown consists of two tubes connected so that the fluctuating currents in the plate circuit of the first tube pass through a transformer and are thereby transferred to the grid circuit of the second tube. The fluctuations of current produced in the plate circuit are always much greater in energy than the charges on the grid. The signals in the first tube are therefore greatly amplified by the second tube.

Amplifiers are arranged to consist of one, two or three *stages,* accordingly as one, two, or three tubes are employed. More than three stages usually distort the signals unless very carefully designed and adjusted.

The amplifier is now commonly used in the modern radio set to increase the signal strength to the point where it will operate a loud speaker. A special form of audion amplifier tube, known as a telephone repeater, is used to strengthen the speech currents over long distance telephone lines.

The next step in the development of radio was the logical one of finding out how to couple up the output plate circuit with the input grid circuit so that the energy of each circuit would react upon the other, thus building up the weak incoming radio oscillations. Or stated differently, to feed back the energy of the output circuit to the input circuit so that the latter would be *regenerated*. This accomplishment, invented by a young man, E. H. Armstrong,* was called the *regenerative* or *feed-back* circuit and coincidentally involved a great discovery. This is that the audion when hooked up with the feed-back circuit and supplied with sufficient energy will develop a sustained alternating current or *sustained local oscillations of its own*. Thus within itself it becomes a source of alternating currents. When arranged in this form it is called an *oscillating* tube and is used as the generator of the oscillatory currents which produce the outgoing waves from the modern broadcasting station. The waves are controlled or *modulated,* that is, made to conform to the vibrations of the voice and music at the transmitting station by an amplifier circuit.

How to Build a Single Tube, Single Control Regenerative Receiver

Receiving sets designed to operate on the 110 volt current can be purchased so cheaply that it does not pay to build one. Not every home has 110 volt current and in the country districts especially a battery operated set is necessary. It is easy to construct, the coils being the only part which it is necessary to make. The circuit is not complicated and since it is only a one tube receiver, the parts to be purchased are not as many or as expensive as they would be if three or four tubes were required. All of the parts may be used again later in the construction of a more elaborate receiver. For these reasons it is ideal for the novice to build as his first attempt in making a radio set.

* Now Prof. Armstrong, Columbia University, N. Y.

The various units which enter into the construction of a radio receiver are manufactured in large quantities and sold at very reasonable prices by radio dealers. A complete set of materials and parts, less the telephone receivers and batteries, as listed below can be purchased for as little as $3.50. It would be very difficult for the amateur not equipped with proper tools and experience to make these parts. The trouble and expense involved in procuring the various materials alone make it uneconomical.

In order to build the one tube set it will be necessary to purchase the following:

PARTS AND MATERIALS

1 21–23 Plate Variable Condenser
1 Three-inch Knob and Dial
1 Grid Condenser
1 Fixed Condenser
1 50 ohm Filament Rheostat
1 No. 30 Vacuum Tube
1 Vacuum Tube Socket
1 Telephone Headset
2 1½ volt No. 6 Dry Cells
1 "B" Battery
1 Three inch dia. Cardboard Tube
6 Binding Posts
60 Feet No. 22 B. & S. gauge Single-Cotton-Covered Wire

The Variable Condenser is for tuning and adjusting the circuit of the receiver to respond to the wave length of the station which it is desired to intercept. Variable condensers consist of two sets of semi-circular plates, mounted in a supporting frame. All of the plates of one set are connected together but insulated from those of the other set. One set is *movable* so that its plates can be rotated between the *fixed* plates without touching them. Rotating the movable plates varies the *capacity* of the condenser. Capacity is a term signifying the *ability of a condenser to store electricity*. The unit of capacity

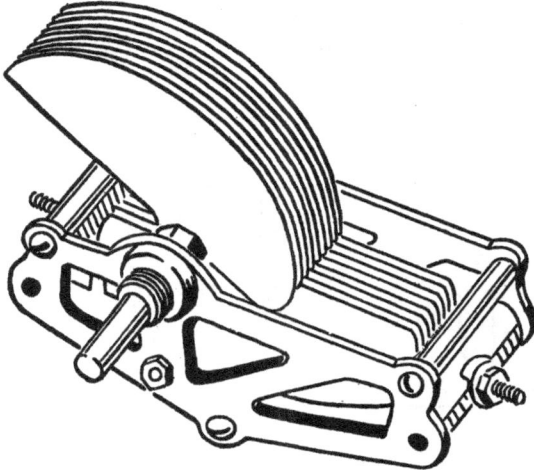

FIG. 208. — A "Straight Line Frequency" Variable Condenser Unit for Tuning and Adjusting the Circuit to the Wave-length of the Station Which It is Desired to Receive.

is a *Farad*, so called after the famous scientist Faraday. A condenser having a capacity of one farad is, however, so large that ordinary condensers are rated in *micro-farads* or thousandths of a farad. The abbreviation for micro-farad is M. F. The variable condenser to be used in making this receiver should have 21–23 plates and a maximum capacity of .0005 M. F. or one-half of one-thousandth of a farad. It should be of the type known as a "straight line frequency" condenser. This characteristic is determined by the shape of the plates which are not exactly semi-circular.

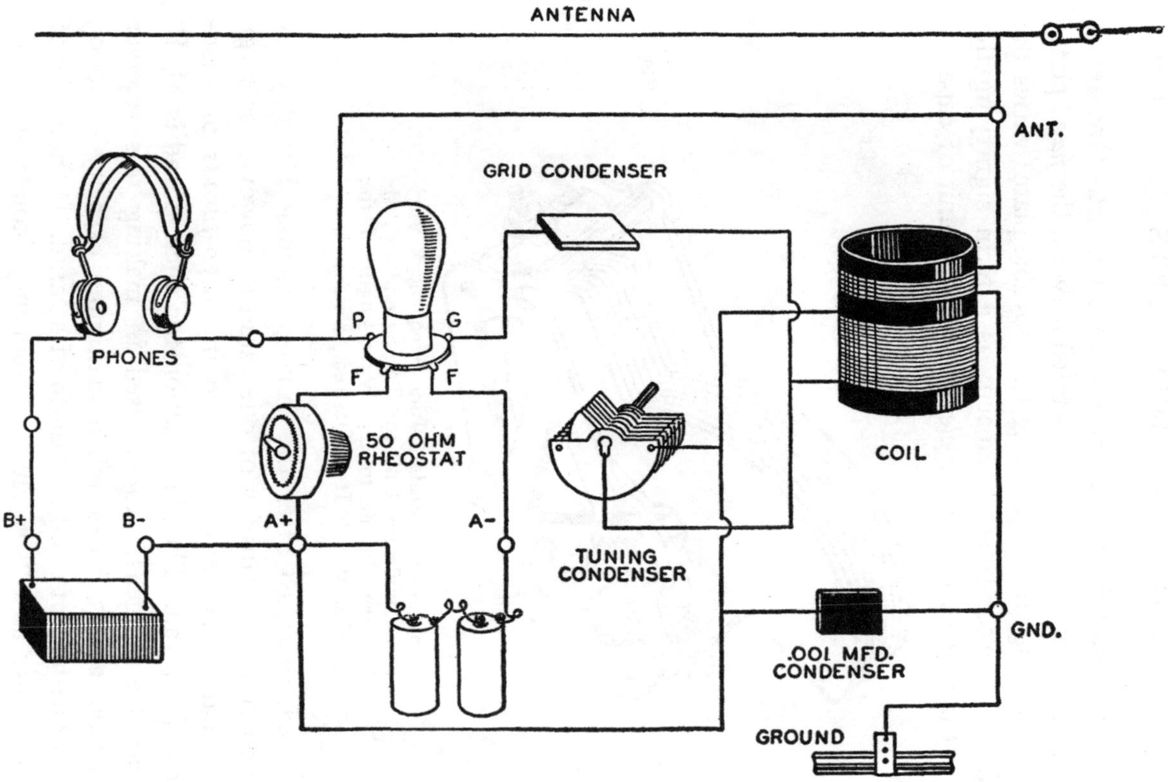

FIG. 209. — The Circuit Diagram for a Single Tube, Single Control Regenerative Receiver.

The Knob and Dial is used to turn the movable plates of the condenser and indicate how far it has been turned. The dial should be three to four inches in diameter. The knob should be provided with a hole to fit the shaft of the condenser and a set screw to fasten it in place.

The Grid Condenser should have a capacity of .00025 M. F. and be provided with clips to fit

The Grid Leak which is a high resistance unit contained in a little glass tube fitted with metal terminals which will

FIG. 210. — The Grid Condenser and Fixed Condenser are alike except in Capacity.

snap into the holes in the clips on the grid condenser. It should have a resistance of two *megohms* or *two million ohms*.

The Fixed Condenser should be one having a capacity of .001 M. F.

The Filament Rheostat regulates the current flowing through the filament to the vacuum tube or audion. It should have a resistance of fifty ohms and be provided with a knob.

The Vacuum Tube is of the type known as No. 30 which is a sensitive detector or amplifier. These tubes are fitted with a standard base having four contact prongs on the bottom. Two contacts lead to the filament and the other two respectively to the plate and grid inside the glass bulb. The side of the base has a small locating pin which fits into a corresponding slot in the socket bringing the proper contact against its corresponding terminal in the socket.

The Socket, needless to say, should be one to fit the tube.

The Telephone Headset should be a standard radio telephone headset consisting of a headband and two receivers having a resistance of 2000 ohms when connected in series.

The Two No. 6 Dry Cells are to furnish current to the filament of the vacuum tube. The battery used for this purpose is called the "A" battery.

The "B" Battery consists of a number of small dry cells built in a unit to furnish 22½ volts. This battery furnishes the current in the plate circuit and produces the sounds in the telephone receivers. Two or three 22½ volt units will be required.

The Binding Posts can be purchased for five cents each, provided with insulated tops marked to indicate to which post

FIG. 211. — A Set of Marked Binding Posts.

the aerial, or antenna, ground, etc., as the case may be, should be connected.

The Coils, of which there are two, are the only part of the set which it will be necessary for you to build yourself. Both are wound on the same cardboard tube. The tube should be three inches in diameter and three and three-eighths inches long. Put it in a warm oven for fifteen or twenty minutes so as to dry it out thoroughly and then give it two or three coats of shellac.

The smaller coil is called the primary and consists of twenty turns of No. 22 B. & S. gauge single-cotton-covered wire. Drill two small holes three-eighths of an inch back from one end of the tube in which to anchor the end of the wire, five or six inches of the latter being left free for connecting later. Wind on twenty turns in a smooth even layer making each turn of wire lie parallel and close to the one next to it. Anchor

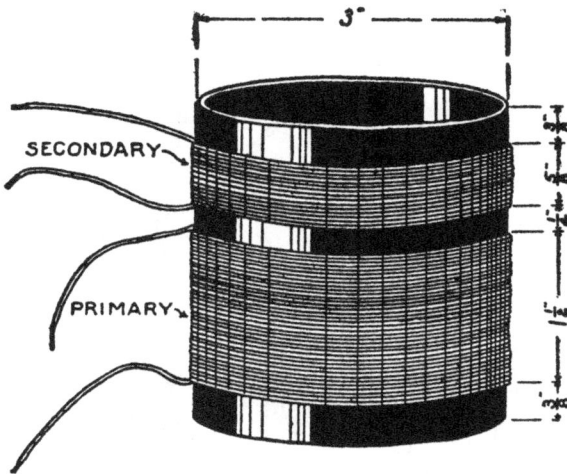

FIG. 212. — Both Coils Are Wound on the Same Cardboard Tube.

the end of the wire by passing it through two small holes drilled for the purpose and leave five or six inches free for connecting.

Start the secondary winding, one-quarter of an inch away from the primary and wind on fifty turns in a smooth, even layer. Make absolutely certain that both coils are wound in the same direction. Otherwise your set will not work. Fasten

SIDE VIEW

5¼ INCHES

1
3
4
2
14-21
5
6
7
13
12
10
9
8
11

PLAN VIEW

GND 21
ANT 20
A BAT + 19
B BAT - 18
A BAT - 17
B BAT + 16
15
14

3
4

5
6

2

7

9
10

8

12

11

13

7 INCHES

8 INCHES

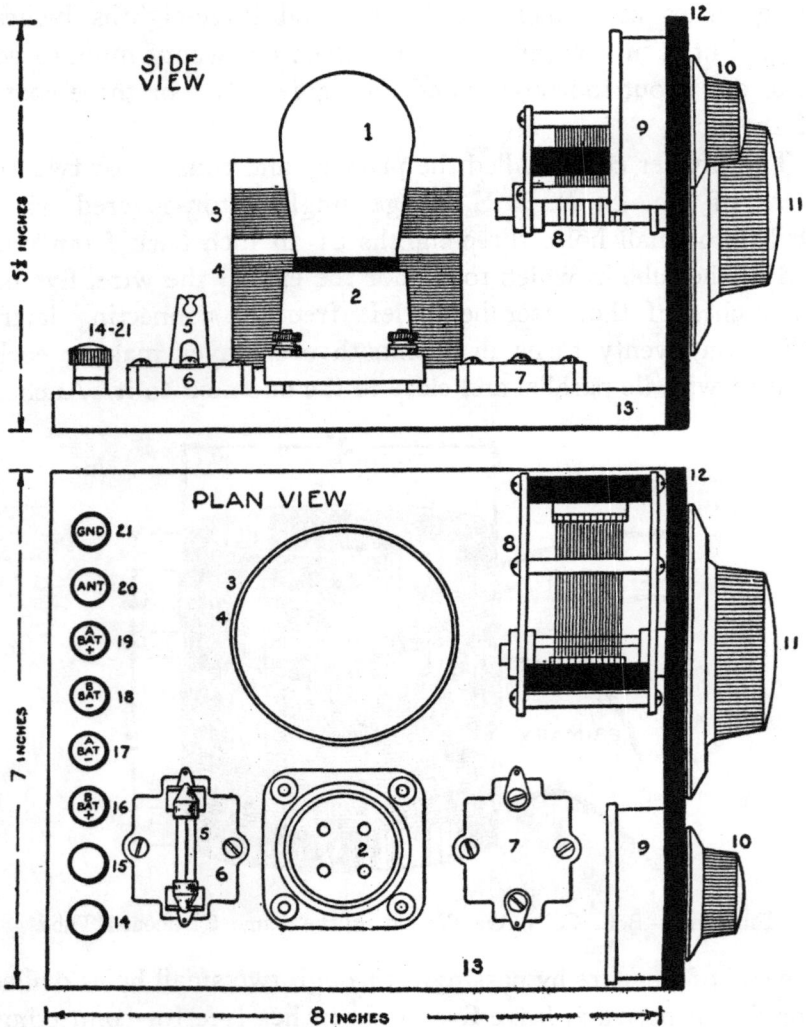

FIG. 213. — Side View and Plan of the Complete Receiver. 1. Vacuum Tube. 2. Tube Socket. 3. Primary Coil. 4. Secondary Coil. 5. Grid Leak. 6. Grid Condenser. 7. Fixed Condenser. 8. Variable Condenser. 9. Filament Rheostat. 10. Rheostat Knob. 11. Dial Knob. 12. Panel. 13. Base. 14–21. Binding Posts.

the ends in the same manner as the primary and leave enough free wire for the connecting later.

Assembling and Connecting the Parts is not difficult but must be done properly if you wish to reap the full benefit of your labors. A neat cabinet will help in giving your work a well finished appearance but if you hope to build another outfit later and use some of the same units it may be well to assemble

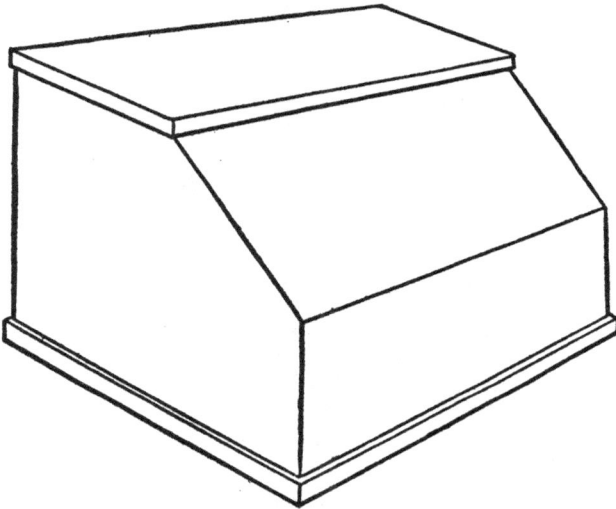

FIG. 214. — A Cabinet with Sloping Front for the Single Tube Receiver.

the parts on a baseboard and panel. For this purpose you will need a piece of well seasoned, dry wood, measuring about seven by eight by one-half an inch and a panel seven by five and one-half inches and one-quarter of an inch thick. The panel may be any one of the composition materials such as Bakelite, Micarta, Condensite, etc., or you may even use a piece of dry wood.

One of the illustrations shows a plan and side view of the different parts mounted on a baseboard and panel. The connecting wires have not been shown for the sake of clearness. All of the parts are numbered and the key given under the sketches so that you will have no difficulty in locating them. The rheostat and the variable condenser are mounted on the panel by means of the screws which are furnished with each

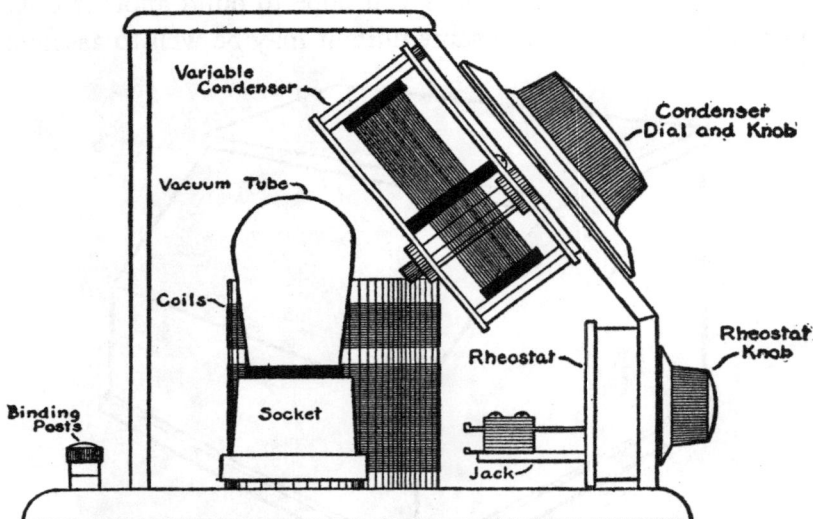

FIG. 215. — Side View of the Receiver Built in the Sloping Front Cabinet. One End Is Shown Open to Expose the Parts and Their Arrangement.

instrument for that purpose. The other parts are screwed down to the baseboard.

Use No. 14 B. & S. gauge bare tinned copper wire and solder all connections. This wire is known as "bus" wire and you can obtain it wherever radio supplies are sold. Use only rosin as the flux in soldering because most soldering pastes contain acid and chemical salts which spread over the surface and cause leakage of the weak radio currents.

If you wish to provide your set with a cabinet, you can give it a very neat appearance by providing one with a sloping

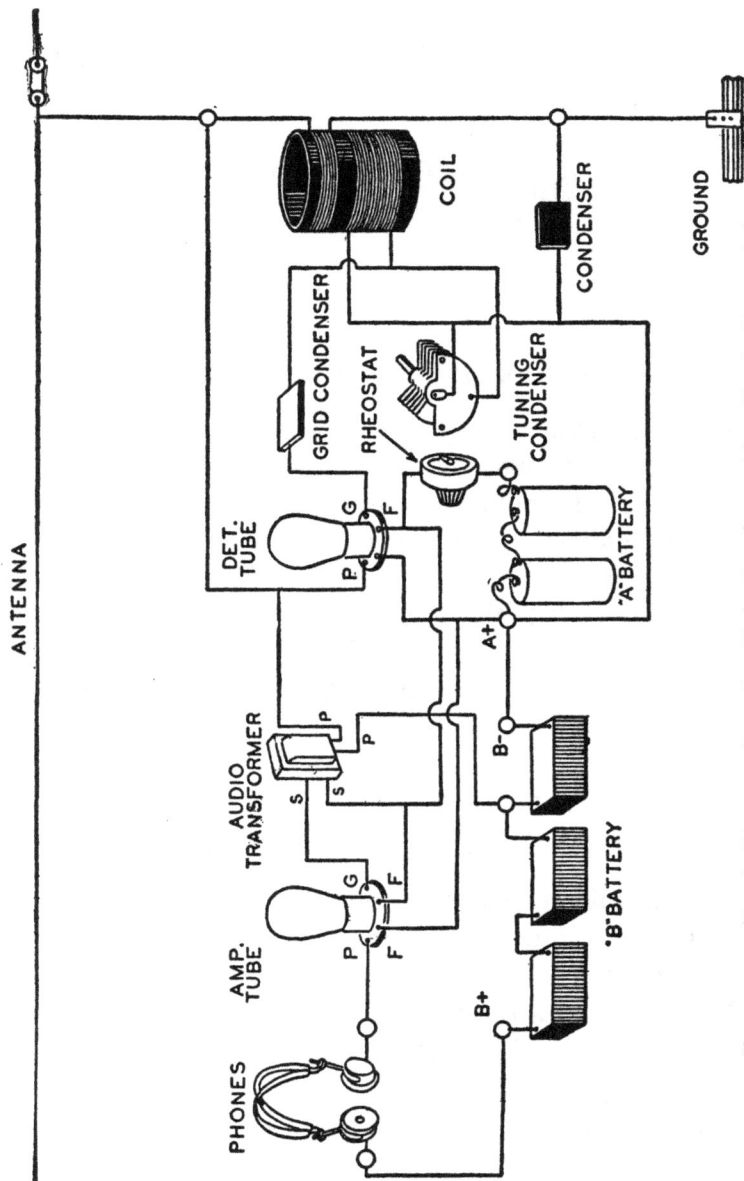

ANTENNA

COIL

GROUND

CONDENSER

GRID CONDENSER

TUNING CONDENSER

RHEOSTAT

DET. TUBE

G

F

P

F

'A' BATTERY

A+

B−

AUDIO TRANSFORMER

P

P

s

s

AMP. TUBE

G

F

P

F

'B' BATTERY

B+

PHONES

Fig. 216. — Circuit Diagram for a One Tube Receiver with a Single Stage of Amplifier.

front. If you wish to provide a "jack" so that the telephone receivers can be connected to the receiver with a plug you will find it very convenient.

WASHER

FIG. 217. — A Jack.

A ONE STAGE AMPLIFIER

A one stage amplifier will greatly increase the strength of the signals which you receive and is very easily added to the set which you have just built. You will require an amplifier tube socket and an audio-frequency transformer and filament rheostat. You will have to make the baseboard for your set slightly larger to accommodate these additional parts. The diagram shows how the connections should be made. Three blocks of 22½ volt "B" battery will be necessary when the amplifier is added.

HOW TO BUILD A FOUR TUBE REGENERATIVE RECEIVER WITH ONE STAGE OF RADIO-FREQUENCY AMPLIFICATION

The vacuum tube may also be used to amplify or strengthen the currents set up in the receiving antenna by the transmitting station, before they pass through the detector tube. When used for this purpose it is a *radio-frequency* amplifier. The modern broadcast receiver usually employs from two to four stages of radio-frequency amplification.

More than one radio-frequency amplifier tube in a set is difficult to control and the inexperienced amateur cannot usually tackle the construction of a five to eight tube receiver with good results unless he buys all of the parts ready made in the form of a complete kit.

Before you spend the money required to buy the parts for a six or eight tube receiver, build the little four tube set described

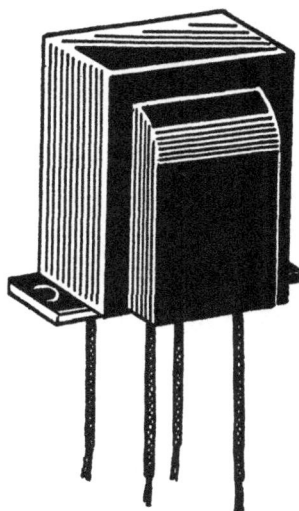

FIG. 218. — An Audiofrequency Transformer.

in the pages which follow. It employs one tube as a radio-frequency amplifier, one as a detector and two as audio-frequency amplifiers. It will give good volume in a loud speaker and is capable of receiving distant signals. The details, as given, are those of a set built by the author which has consistently received the signals of a broadcasting station located in California, during the winter months night after night. The receiver was located in New Jersey.

In order to build this four tube set, it will be necessary to purchase the following

Parts and Materials

1 .005 M. F. Straight Line Frequency Variable Condenser
1 Four-inch Knob and Dial
1 .00025 M. F. Grid Condenser
1 Three Megohm Grid Leak
1 .002 M. F. Fixed Condenser
1 .00025 M. F. Fixed Condenser
1 Filament Rheostat
4 Type 30 Vacuum Tubes
4 Vacuum Tube Sockets
2 Audio-Frequency Transformers
2 Single Circuit Jacks
1 Single Pole Battery Switch
1 Single Pole Double-throw Battery Switch
2 No. 6 Dry Cells
2 45-volt "B" Batteries
8 Binding Posts
  Miscellaneous Magnet Wire, etc., as described below
1 Loud Speaker

The round black circles in the circuit diagram indicate the position of the various binding posts.

It will be noticed that the circuit shows several coils marked *A, B, C, D* and *E*. Coils *A* and *B* are a radio-frequency transformer and *C, D* and *E* a device which we will call a "volume control." These two devices are to be made by the experimenter and are described below.

The Radio-Frequency Transformer consists of two separate coils, marked *A* and *B*, each one mounted on a cardboard tube.

The larger tube has an outside diameter of two and one-quarter inches and is two and one-eighth inches long. The coil,

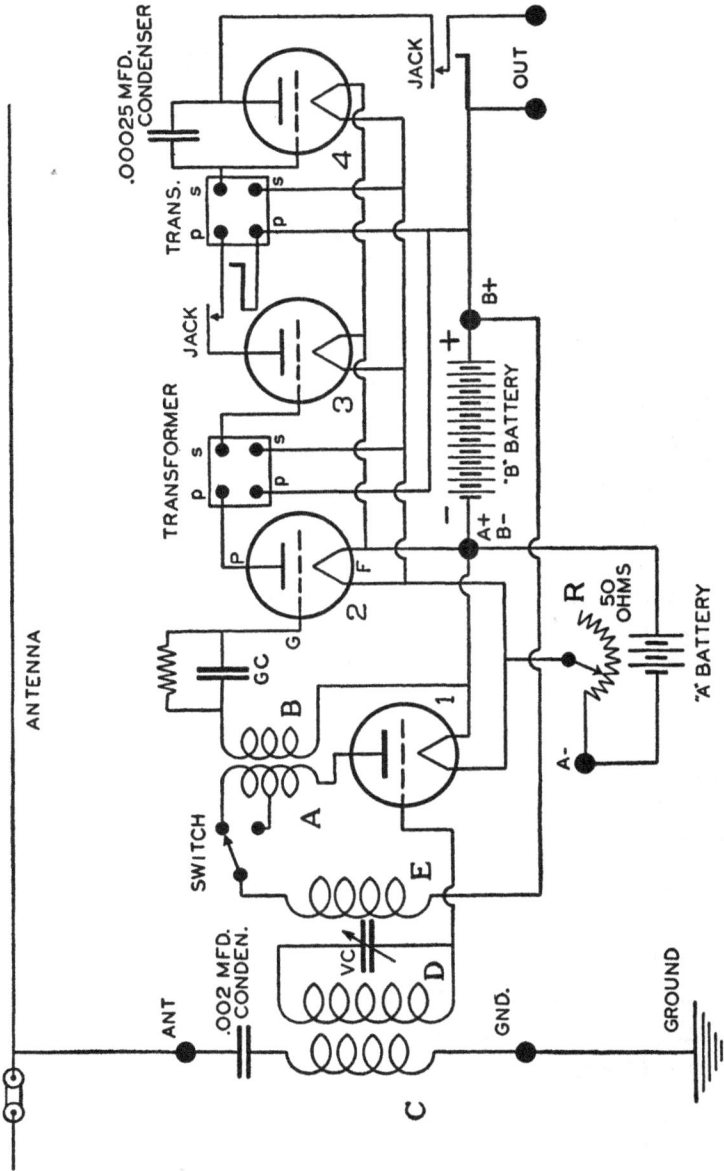

FIG. 219.—A Diagram of the Circuit of the Four Tube Receiver: 1. Radio-frequency Amplifier Tube. 2. Detector Tube. 3. First Audio Amplifier Tube. 4. Second Audio Amplifier Tube. *A* and *B*. Radio-frequency Transformer. *C*, *D* and *E*. "Volume" Control. *GC*. Grid Condenser. *VC*. Variable Condenser. *R*. Rheostat. *P*, *G* and *F* indicate the Plate, Grid and Filament of a Tube. *P* and *S* indicate the Primary and Secondary Terminals of a Transformer.

A, consists of ninety-two turns of No. 30 B. & S. gauge double-cotton-covered magnet wire, wound on in a single, smooth layer and then varnished. A "tap" or connection is led off from the thirty-seventh turn.

The smaller tube is two inches in diameter and one and one-half inches long. The coil, *B,* consists of seventy-eight turns

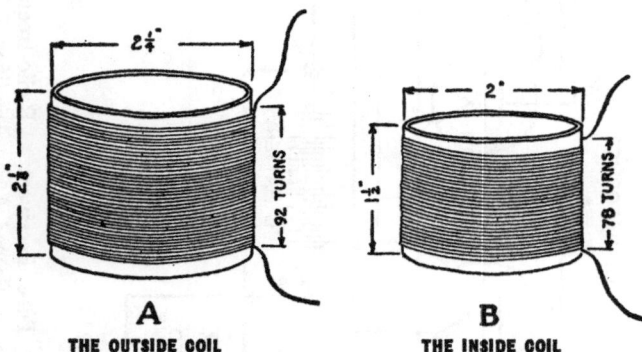

Fig. 219 A. — Details of the Coils Which Form the Radio Frequency Transformer.

of No. 35 B. and S. gauge double-cotton-covered magnet wire wound on in a single smooth layer and then varnished.

Both tubes should have a wall about one-sixteenth of an inch thick. Cardboard or paper tubes which have been dried and shellacked may be used. Bakelite or micarta make a better mechanical job.

Both coils should be wound in the same direction. The *B* is slipped inside of the larger coil, *A,* and fastened there with two or three strips of varnished paper.

The Volume Control consists of three coils marked *C, D* and *E* in the diagram. *E* is movable and arranged so that it can be rotated inside of *C* and *D.* The rotating coil is wound on a tube one and five-eighths of an inch in diameter and one inch long. This tube is mounted on a shaft which passes through the tube supporting the windings C and D, at right angles to the axis.

The coil E consists of fourteen turns of No. 25 B. and S. gauge double-cotton-covered magnet wire. Seven turns are wound on each side of the shaft.

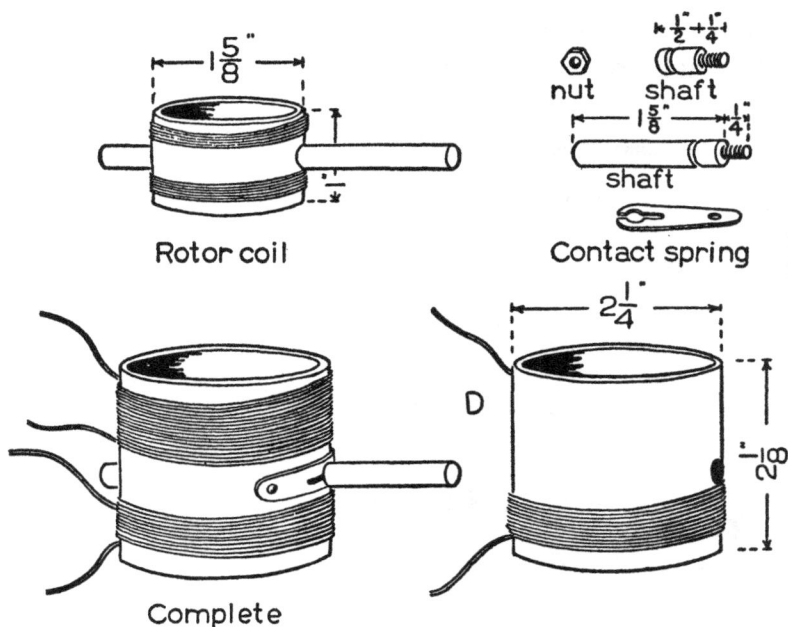

FIG. 220. — Details of the Coils which form the Radio-frequency Transformer.

The coil D consists of sixty-five turns of No. 34 B. and S. gauge double-cotton-covered wire wound on a tube two and one-quarter inches in diameter and two and one-eighth inches long. Thirty-seven turns are wound on that portion above where the shaft of the rotating coil passes through and twenty-eight turns below.

Coil *C* consists of five turns of No. 14 B. and S. gauge double-cotton-covered magnet wire wound directly over the lower portion of *D*.

The Jacks shown in the circuit diagram and included in the list of materials make it possible to use either one or two stages of audio-frequency amplification at will. The cord attached to the loud speaker is connected to a plug which will slip into either one of the jacks.

The single pole battery switch marked "F.S." in the diagram

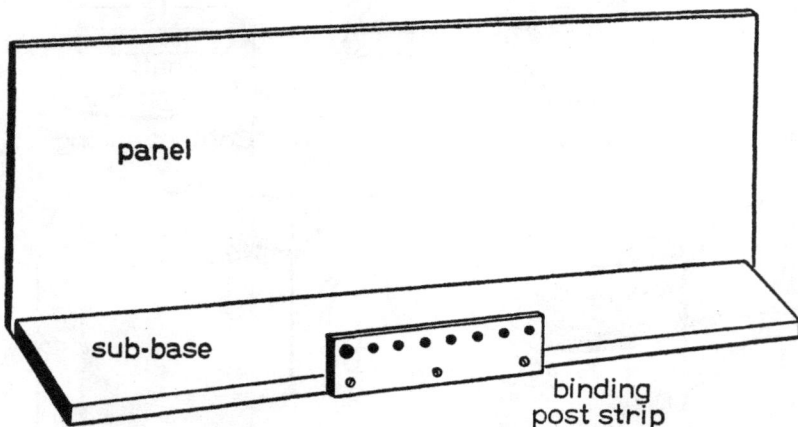

panel

sub-base

binding
post strip

Fig. 221. — Back View showing how the Panel is mounted on wooden Sub-base.

is used to disconnect the filament lighting current when the receiver is not in use.

The single pole, double throw switch, "W.S.," connects either a portion or all of coil *A*, in the circuit at will, accordingly as the longer or shorter broadcasting waves are to be received.

The Loud Speaker to be used with the set should be of the magnetic type and not a dynamic speaker. The 5-inch or 6-inch is the proper size. The cost will be approximately $1.00.

Assembling the Receiver. The illustrations will give an idea of how to arrange a panel and cabinet so that the parts may be efficiently mounted and housed. The panel is seven by nineteen inches and three-sixteenths of an inch thick.

The rheostat, variable condenser, jacks and switches are mounted on the panel. The shaft from the rotating coil, *E,* projects through the front of the panel and is fitted with a knob and a pointer, marked "Volume."

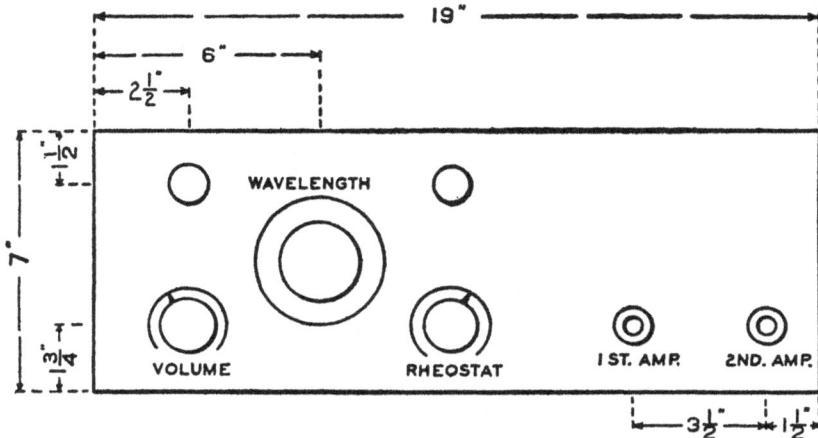

FIG. 222. — A Front View of the Panel With All the Parts in Position.

The rest of the parts are mounted on a sub-base attached to the back of the panel.

The binding posts are mounted on a strip of micarta or bakelite at the back of the sub-base.

A slot cut in the back of the cabinet allows the binding posts to project through so that they can be conveniently reached when making connections.

The cabinet should be constructed so that the top opens and the tubes can be put in or taken out or the set examined by raising the lid.

The front edges of the sides and the top of the cabinet are rabbeted to fit the panel.

The receiver is tuned by adjusting the "Volume" control and the variable condenser. The adjustment of the rheostat

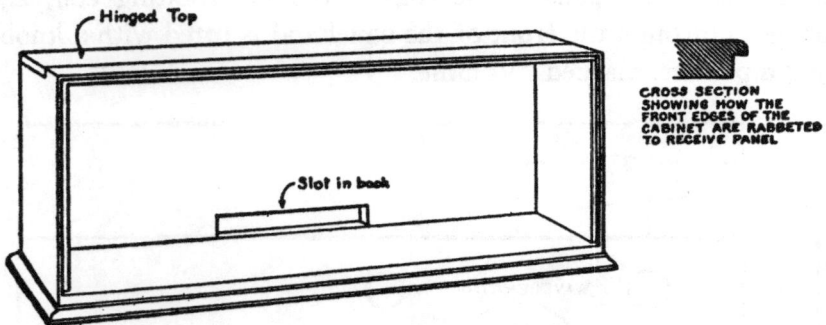

CROSS SECTION SHOWING HOW THE FRONT EDGES OF THE CABINET ARE RABBETED TO RECEIVE PANEL

FIG. 223. — Details of the Cabinet.

will affect the sensitivity of the apparatus. It does not need to be changed for each station but only as the condition of the dry cells varies.

Tuning the Receiver is a comparatively simple operation. One or two evenings' practice will make almost any one an expert.

Insert the plug connected with the loud speaker cord in the second stage amplifier jack. Set the rheostat about halfway on the semi-circle and turn on the filament lighting switch. Set the "Tuning" dial at about 50 degrees. Turn the "Volume" control until there is a "cluck" or hissing sound in the loud speaker and while keeping the volume control adjusted so as to maintain this condition, slowly rotate the tuning dial between 30 and 100 degrees.

If there is a broadcasting station in operation within the range of your receiver, it will cause a low whistling sound to be heard in the loud speaker. As soon as this whistle is heard, turn the volume control back until the whistle stops. Then turn it forward again very slowly to the point where the whistling just starts to take place and at the same time vary the tuning dial back and forth one or two degrees until the signal is clear and of sufficient volume.

The proper setting should be "logged" or noted for each station. Once the setting of any station on the tuning dial is known, the proper method of tuning is as follows:

Set the tuning dial at the exact point where the station was previously received and slowly turn the volume control until the signal is strong and clear.

The proper position of the "wavelength switch" will vary accordingly as the station desired tunes in on the upper or the lower part of the tuning dial. When through using the receiver be certain to turn the filament switch so as to extinguish the tubes.

CHAPTER XVI

AN EXPERIMENTAL "WIRELESS" TELEPHONE

THE device described in the following pages is easy to make and arrange, and will serve for some interesting experiments.

It is of no practical value as a telephone, because the distance over which it will transmit speech is limited to from 250 to 300 feet. If you have a pal who lives across the street and within the distance named above, it is possible for you to remain in your own rooms and talk to each other without any connecting wires.

The instruments operate by magnetic induction. It has already been explained how it is possible for the current in the primary of an induction coil to induce a current in the secondary coil, even though the two are not electrically connected. This type of wireless telephone really consists of an induction coil in which the two windings are widely separated.

Suppose that two coils of wire are connected as in the illustration so that one coil, A, is connected in series with a set of batteries and a telegraph key. The terminals of the other coil, B, are connected to a telephone receiver. The coils are placed parallel to each other and a few inches apart. If the key is pressed so that the battery current may flow through the coil, A, it will create a magnetic field, and lines of force will be set up in the immediate vicinity. The lines of force will pass through the coil, B, and induce in it a current of electricity which will cause a sound like a click to be heard in the telephone receiver.

If a telephone transmitter is substituted for the key and words are spoken into it, the current passing through the coil from the battery will vary with each vibration of the voice and the words will be distinctly repeated by the receiver connected to *B*.

This experiment may be tried by any boy with the equipment he probably has already around his shop. Twenty-five

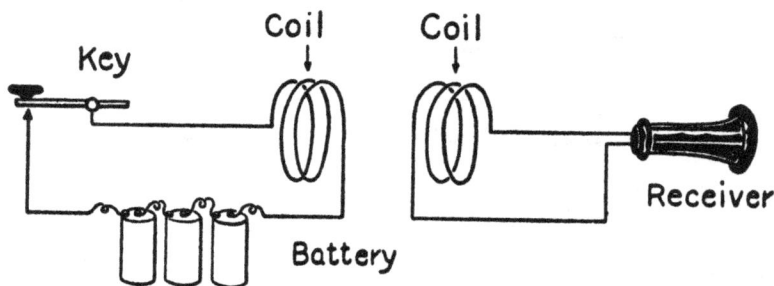

Fig. 224. — A Simple Arrangement Showing the Inductive Action Between Two Coils.

to thirty turns of wire wound around a cardboard tube five or six inches in diameter will serve as a coil. Two such coils, an ordinary telephone transmitter, a telephone receiver and a couple of dry cells are all that is required.

The diagram in the accompanying illustration shows how the apparatus is arranged. The coils may be used several inches apart and the voice will be clearly heard in the receiver.

Such an outfit is, however, only experimental, and if it is desired to make a practical set, the coils, etc., must be much larger in diameter and contain a greater number of turns.

Larger coils are made by first drawing a circle four feet in diameter on the floor of the "shop." Then drive a number of small nails around the circumference about four inches apart.

Wind two and one-half pounds of No. 20 B. & S. gauge cotton-covered magnet wire around the circumference of the circle. The wire should form at least sixty complete turns. About one foot should be left at each end to establish connections with. After winding, the coil should be tied about every six inches with a small piece of string so that it will

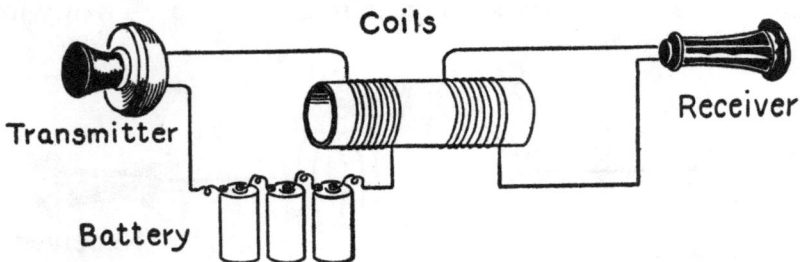

FIG. 225. — A "Wireless" Telephone Experiment. Speech directed into the Transmitter can be heard in the Telephone Receiver although there is no direct electrical connection between the two.

retain its shape and not come apart. The nails are then pulled out so that the coil may be removed.

The coil may be used just as it is for experimental purposes, but if it is intended for any amount of handling it is wise to procure a large hoop such as girls use for rolling along the sidewalk, and make the coil the same diameter as the hoop so that upon completion they may be firmly bound together with some insulating tape. Two binding-posts may then be fastened to the hoop and the terminals of the coil connected to them.

Two such coils are required for a complete wireless telephone system, one to be located at each station.

It is also necessary to make a double-contact strap-key. Such a key is easily built out of a few screws and some sheet-brass. The illustration shows the various parts and construction so clearly that no detailed explanation is necessary.

The telephone transmitter and the telephone receiver required for this experiment must be very sensitive, and it is hardly possible for the young experimenter to build one which will be satisfactory. They can be secured from a second-hand tele-

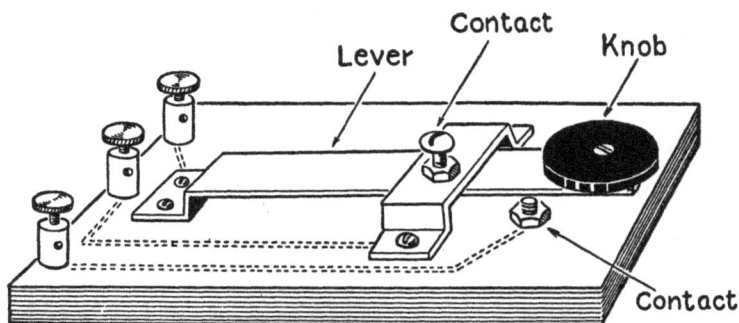

FIG. 226. — A Double-Contact Strap-Key for use with the electromagnetic Wireless Telephone.

phone or purchased at an electrical supply house. The transmitter should be of the "long distance" type. An 80-ohm receiver will serve the purpose, but it is much more satisfactory to use a pair of 1000-ohm radio receivers.

A battery capable of delivering about 12 volts is required.

The apparatus is connected so that when the key is pressed, the coil is connected to the battery and the telephone transmitter. If words are then spoken into the transmitter they will vary the amount of current flowing and the magnetic field which is set up in the neighborhood of the coil will induce currents in the coil at the other station, provided that it is not too far away, and cause the words to be reproduced in the telephone receiver.

When the key is released it will connect with the upper contact and place the telephone receiver in the circuit so that your pal at the other station can answer your message by pressing his key and talking into his transmitter.

The best plan is to mount each of the coils upon a tripod and experiment by placing them close together at first and grad-

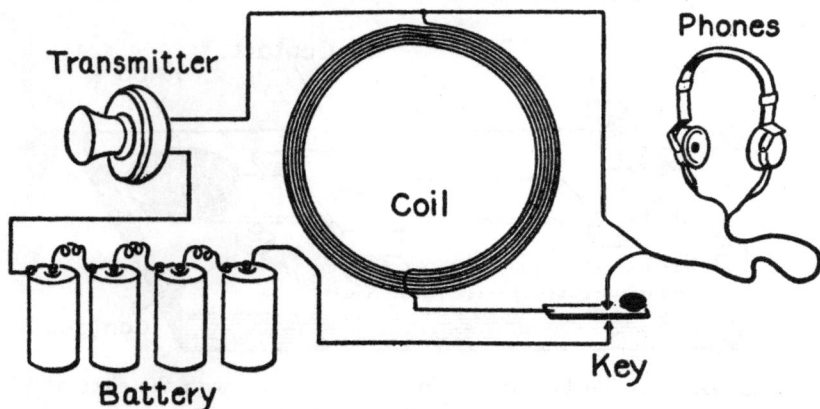

FIG. 227. — The Circuit of the "Wireless" Telephone. When the Key is up, the Phones are connected to the Coil. When the Key is pressed, the Transmitter and Battery are connected to the Coil.

ually moving them apart until the maximum distance that the apparatus will work is discovered.

Be very careful to keep the two coils exactly parallel.

Do not hold the key down any longer than is absolutely necessary, or the telephone transmitter will become hot.

By making the coils six feet in diameter and placing from 200 to 400 turns of wire in each coil you can transmit speech 300 feet or more. The coil may be mounted on the wall of your shop in a position where it will be parallel to one located in your friend's house.

The success of a telephone system of this sort lies in making the coils of large diameter and many turns, in keeping the coils parallel, using a sensitive transmitter and receiver, and in employing a strong battery. Storage cells are the best for the purpose.

CHAPTER XVII

ELECTRIC MOTORS

THE first American patentee and builder of an electric motor was Thomas Davenport. The father of Davenport died when his son was only ten years old. This resulted in the young inventor being apprenticed to the blacksmith's trade at the age of fourteen.

Some years later, after having thoroughly learned his trade, young Davenport married a beautiful girl of seventeen, named Emily Goss, and settled in the town of Brandon, Vermont, as an independent working blacksmith.

About this time Joseph Henry invented the electromagnet. Davenport heard of this wonderful "galvanic magnet" which it was rumored would lift a blacksmith's anvil. This was his undoing, for never again was he to know peace of mind but was destined to always be a seeker after some elusive scientific "will-o'-the-wisp." Although many times he needed iron for his shop, the greater part of his money was spent in making electromagnets and batteries.

In those days insulated wire could not be purchased, and any one wishing insulated wire had to buy bare wire and insulate it himself. It was then supposed by scientists that silk was the only suitable material for insulating wire and so Davenport's young wife cut her silk wedding gown into narrow strips and with them wound the coils of the first electric motor.

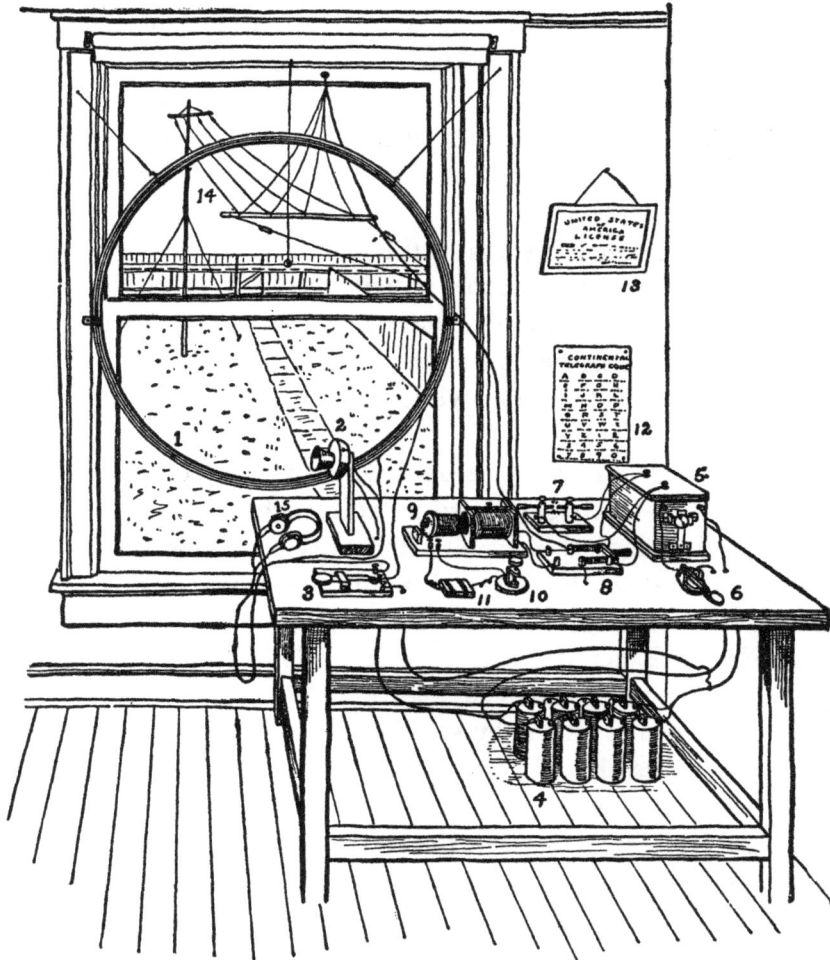

FIG. 228. — An Amateur Wireless Telegraph Station twenty-five years ago when Spark Coils were used to send Messages. The coil of an electro-magnetic "Wireless" Telephone (1) hangs in the window. In the yard is an old-fashioned 4-wire Antenna (14). 2. The Telephone Transmitter. 3. Double-Contact Strap Key. 4. The Battery. 5. Spark Coil. 6. Key. 7. Spark-Gap. 8. Aerial Switch. 9. Loose Coupler. 10. Detector. 11. Fixed Condenser. 12. Code Chart. 13. Amateur License. 14. Antenna. 15. Tele-phone Receivers.

Continuing his experiments in spite of almost insurmountable difficulties and making many sacrifices which were equally shared by his family, Davenport was enabled to make a trip to Washington in 1835 for the purpose of taking out a patent. His errand was fruitless, however, and he was obliged to return home penniless.

Nothing daunted, he made the second and third trip and finally secured his memorable patent, the first of the long line of electric-motor patents that have made possible both the electric locomotive that hauls its long train so swiftly and silently, and the whirring little fan which stirs up a breeze during the hot and sultry days.

These are a few of the reasons why a modest country blacksmith, in turn an inventor and an editor, through perseverance in struggling against adversity and poverty succeeded in placing his name on the list which will be deservedly immortal among the scientists and engineers of the world.

A Simple Electric Motor can be made in fifteen minutes by following the plan shown in Figure 229.

The armature is made by sticking a pin in each end of a long cork. The pins should be as nearly central as it is possible to make them, so that when the cork is revolved upon them it will not wabble. The pins form the shaft or spindle of the motor. Wind the cork with about ten feet of fine magnet wire (Nos. 28–32 B. & S. gauge) as shown in the illustration, winding an equal number of turns on each side of the two pins.

When this is finished, fasten the wire securely to the cork by binding it with thread.

Bend the two free ends (the starting and the finishing end) down at right angles and parallel to the shaft so as to form two commutator sections as shown in the upper left-right corner of Figure 229. Cut them off so that they only project about

three-eighths of an inch. Bare the ends of the wire and clean
them with a piece of fine emery paper or sandpaper.

The bearings are made by driving two pins into a pair of
corks so that the pins cross each other.

They must not be at too sharp an angle, or when the arma-

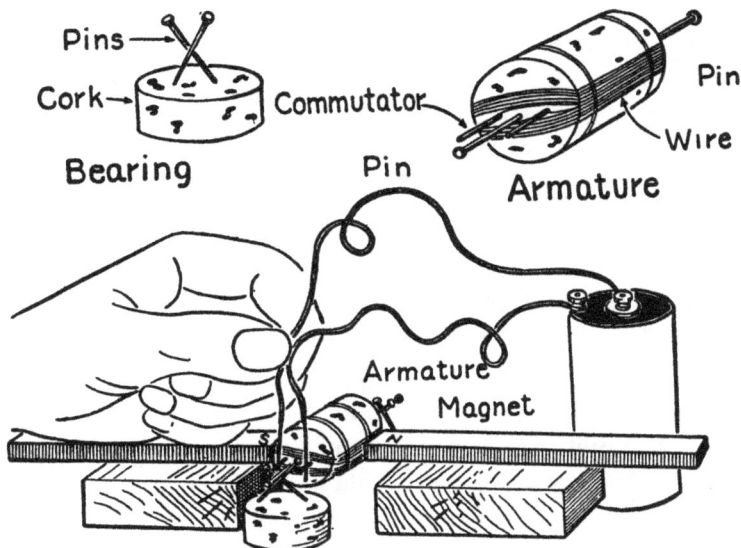

FIG. 229. — A Simple Electric Motor Which Can Be Made in
Fifteen Minutes.

ture is placed in position, the friction of the shaft will be so
great that it may not revolve.

The motor is assembled by placing the armature in the bearings and arranging two bar magnets on either side of the armature. The magnets may be laid on small blocks of wood and
should be so close to the armature that the latter just clears
when it is spun around by hand. The north pole of one magnet should be next to the armature and the south pole of the
other, opposite.

Connect two No. 26 B. & S. gauge wires about twelve inches long to a dry cell. Bare the ends of the wires for about. an inch and one half.

With the ends of the two wires between the forefinger and thumb bend them out, so that when the armature is revolved they can be made just to touch the ends of the wire on the armature, or the "commutator sections," as they are marked in the drawing.

Give the armature a twist so as to start it spinning, and hold the long wires in the hand so that they make contact with the commutator as it revolves.

Very light pressure should be used. If you press too hard, you will prevent the armature from revolving, while, on the other hand, if you do not press hard enough, the wires will not make good contact.

The armature will run in only one direction. If you start it in the right direction and hold the wires properly, it will continue to revolve at a high rate of speed.

If carefully made, this little motor will reward its maker by running very nicely. Although it is of the utmost simplicity it demonstrates the same fundamental principles which are employed in real electric motors.

The Simplex Motor is a little toy which can be made in a couple of hours, and when finished it will make an instructive model.

As a motor itself, it is not very efficient, for the amount of iron used in its construction is necessarily small. The advantage of this particular type of motor and the method of making it is that it demonstrates the principle that is used in larger machines.

The field of the motor is of the type known as the "simplex," while the armature is the "Siemens H" or two-pole type.

The field and the armature are cut from ordinary tin-plated iron such as is used in the manufacture of tin cans and cracker-boxes.

The simplest method of securing good flat material is to get some old scrap from a plumbing shop. An old cocoa tin or baking-powder can may, however, be cut up and flattened and will then serve the purpose almost as well.

The Armature. Two strips of tin, three-eighths of an inch by one and one-half inches, are cut to form the armature. They

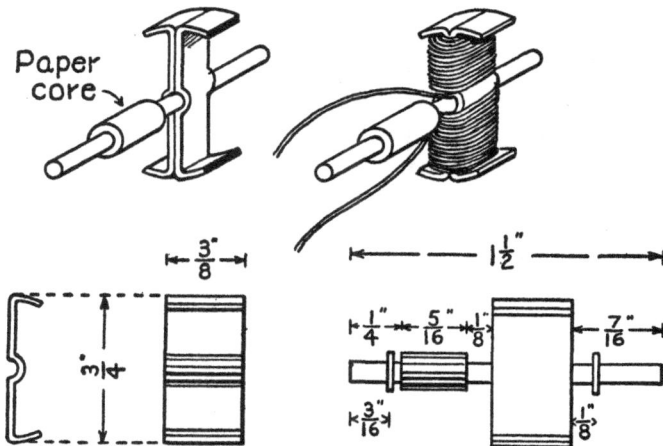

FIG. 230. — Details of the Armature of the Simplex Motor.

are slightly longer than will actually be necessary, but are cut to length after the finish of the bending operations. Mark a line carefully across the center of each strip. Then, taking care to keep the shape symmetrical so that both pieces are exactly alike, bend them into the shape shown in Figure 230. The small bend in the center is most easily made by bending the strip over a knitting-needle and then bending it back to the required extent.

A piece of knitting-needle one and one-half inches long is required for the shaft. Bind the two halves of the armature together in the position shown in Figure 230. Bind them with a piece of iron wire and solder them together. The wire should be removed after they are soldered.

The Field Magnet is made by first cutting out a strip of tin one-half by four and then bending it into shape. The easiest way of doing this with accuracy is to cut out a piece of wood

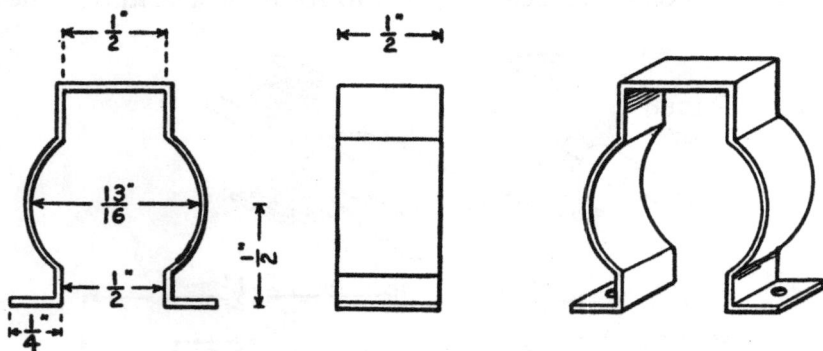

FIG. 231. — The Field Frame is formed out of Sheet Metal.

as a form, and bend the tin over the form. The dimensions shown in the illustration should be used as a guide for the form.

Two small holes should be bored in the feet of the field magnet to receive No. 3 wood screws, which fasten the field to the base.

The Bearings are easily made by cutting from sheet-tin. Two small washers, serving as collars, should be soldered to the shaft.

The Commutator Core is formed from a strip of paper five-sixteenths of an inch wide and about five inches long. It should be given a coat of shellac on one side and allowed to get sticky. The strip is then wrapped around the shaft until its diameter is three-sixteenths of an inch.

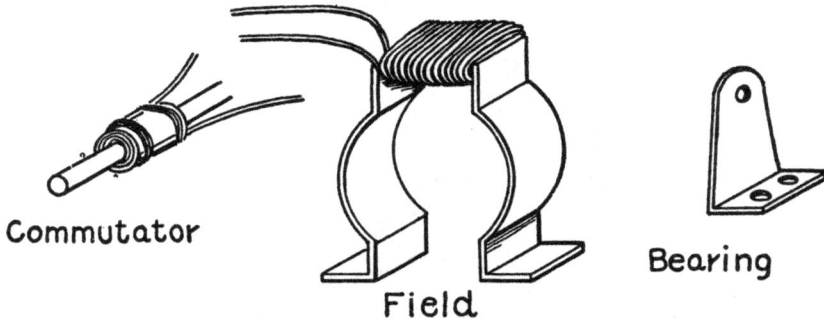

Commutator   Bearing

Field

FIG. 232. — The Commutator Field and Bearing.

The Base is cut from any ordinary piece of wood and is in the form of a block about two by one and one-half by one-half inch.

Assembling the Motor. Those portions of the field and armature which are to be wound must be prepared by covering with paper. Cut a strip of paper one-half inch wide and one and one-eighth of an inch long and give it a coat of shellac on one side. As soon as it becomes sticky, wrap it around the top bar of the field magnet. The armature is insulated in exactly the same way, taking care that the paper covers the entire flat portion.

The field and armature are now ready for winding. It is necessary to take proper precautions to prevent the first turn from slipping out of place.

This is accomplished by looping a small piece of tape or cord over it. The next two turns are then taken over the ends of the loop so as to embed them. Wind on three layers of wire and when in the middle of the fourth layer embed the ends of another loop, which may be used at the end of the fourth layer to fasten the end so that it will not unwind. After the winding is finished, give it a coat of shellac.

The winding of the armature is somewhat more difficult.

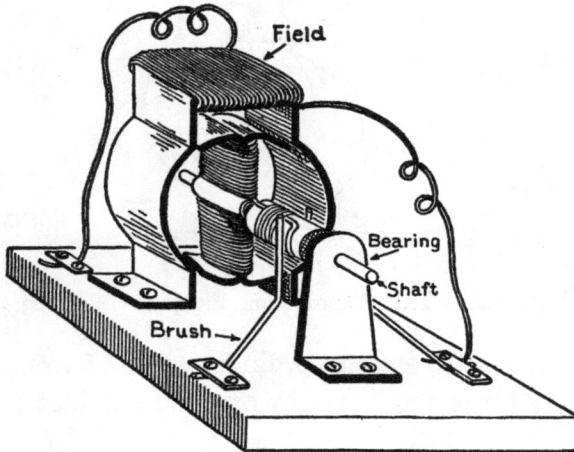

FIG. 233. — The Complete Simplex Motor.

The wire used for winding both the armature and the field should be No. 25 or No. 26 B. & S. gauge double-cotton-covered.

In order to wind the armature, cut off about five feet of wire and double it back to find the center. Then place the wire diagonally across the center of the armature so that there is an equal length on both sides. Place a piece of paper under the wire at the crossing point to insulate it. Then, using one end of the wire, wind four layers on half of the armature. Tie the end down with a piece of thread and wind on the other half.

The ends of the wire are cut and scraped to form the commutator segments.

Bend the wires as shown so that they will fit closely to the paper core. Bind them tightly into position with some silk thread. Use care so that the two wires do not touch each other. Cut the free ends of the wires off close to the core.

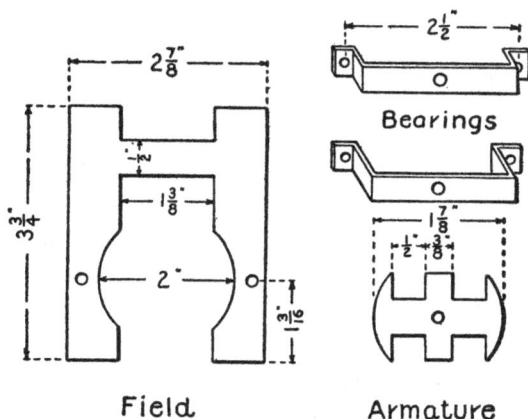

Field   Armature

FIG. 234. — Details of the Field, Bearings and Armature for the Motor Shown in Fig. 236.

When finished, the relative positions of the armature and the commutator should be as shown in Figure 233.

The brushes are made by flattening a piece of wire by a few light hammer blows.

The brushes are fastened under a small clamp formed by a strip of tin held down at each end with a wood screw. They can be adjusted to the best advantage only under actual working conditions when the current is passing through the motor. One or two dry cells should be sufficient to operate the motor.

One end of the field winding is connected to one of the brushes. The other brush and the other end of the field form the terminals to which the battery is connected.

The motor, being of the two-pole armature type, must be started when the current is turned on by giving it a twist with the fingers.

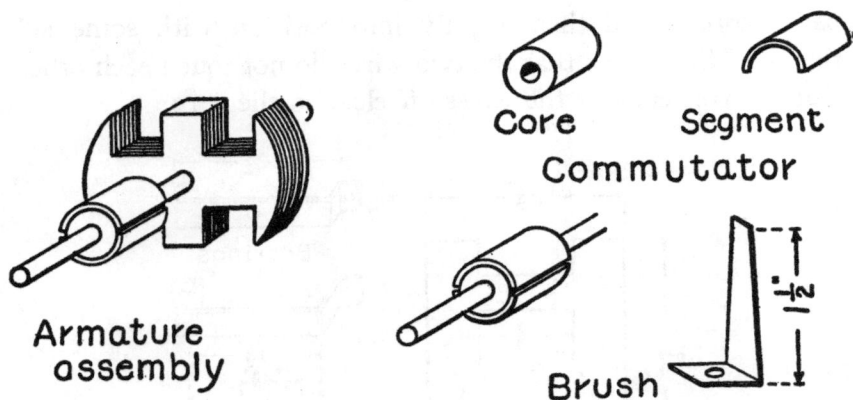

Core   Segment

Commutator

Armature
assembly

Brush

FIG. 235. — The Armature Commutator and Brush.

A Larger Motor may be built in somewhat the same manner as the one just described by cutting armature and field out of sheet tin. It can be made more substantial if it is built up out of laminations and not bent into shape like the little simplex motor just described.

Lay out an armature disk and a field lamination on a sheet of tin in accordance with the dimensions and pattern shown in Figure 234. These pieces are used as patterns for laying out the rest of the laminations.

Place them on some thin sheet-iron and trace the outline with a sharp-pointed darning-needle. Cut a sufficient number of pieces of each pattern to form a pile three-quarters of an inch thick.

Four laminations for the field should be cut with extensions shown by the dotted lines. They are bent out at right angles for mounting the motor and holding it upright.

Assemble the armature and field by piling the pieces on top of each other and truing them up. Enough laminations should be used to form a pile three-quarters of an inch thick when piled up and clamped tightly.

File off any burrs and rough edges and then bind the laminations together with some string to hold them until wound. Then remove the string.

Wrap a couple of layers of paper around those portions of the armature and field which are liable to come into contact with the iron. Five or six layers of No. 18 B. & S. gauge

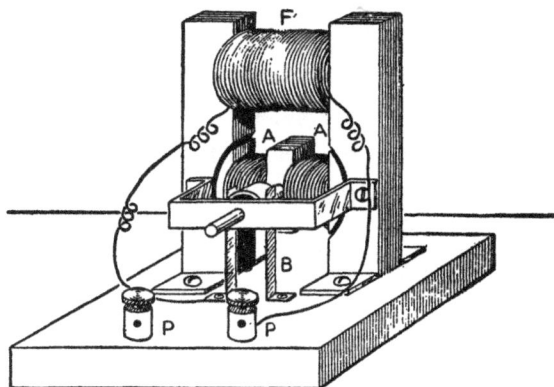

Fig. 236. — The Completed Motor.

double-cotton-covered magnet wire are sufficient to form the field coil.

The armature is wound with three or four layers of wire of the same size.

The commutator is made out of a circular piece of hard wood or fiber, fitted with segments cut out of thin sheet-copper. The segments may be fastened to the core with thick shellac or some melted sealing-wax. The ends may be bound down tightly by wrapping with silk thread.

The brushes are cut out of thin sheet-copper similar to that used for the commutator segments.

The bearings are strips of heavy sheet-brass bent into the shape shown. They are mounted by passing a nail through the holes in the ends and through the holes in the field and then riveting the ends over.

If desirable, a small pulley may be fitted to the shaft and the motor used to run small mechanical toys. If it is properly constructed, two or three dry cells will furnish sufficient current to run the motor at high speed.

How to Build an Electric Engine

An electric engine is really a form of electric motor but differs from the most common form of motor in that the armature, instead of revolving, moves back and forth, like the piston of a steam or gasoline engine. Electric engines are not as efficient as electric motors from the standpoint of the amount of

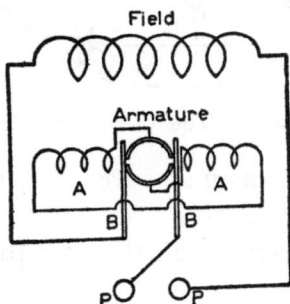

Fig. 237.—Diagram showing how to connect the Motor.

power delivered in proportion to the current used, but they make very interesting models.

Various forms of electric engine were made before the first practical electric motor was invented. They amounted to little more than curiosities and could only be used where the expense of electric current was not to be regarded.

The engine illustrated is of the double action type. It is provided with two electromagnets arranged so that one pulls the armature forward and the other pulls it back. The motion of the armature is transmitted to the shaft by means of a connecting rod and crank. It is very simple to build and the design is such that it will operate equally well whether it is made large or small. If you do not happen to have all the necessary materials to build an engine according to the dimensions shown in the drawings you can make it almost any size and it will work equally well.

All the various parts have been marked so that you can easily identify them in the other drawings. It is well to study this illustration carefully first so that you will understand just how all the parts are arranged.

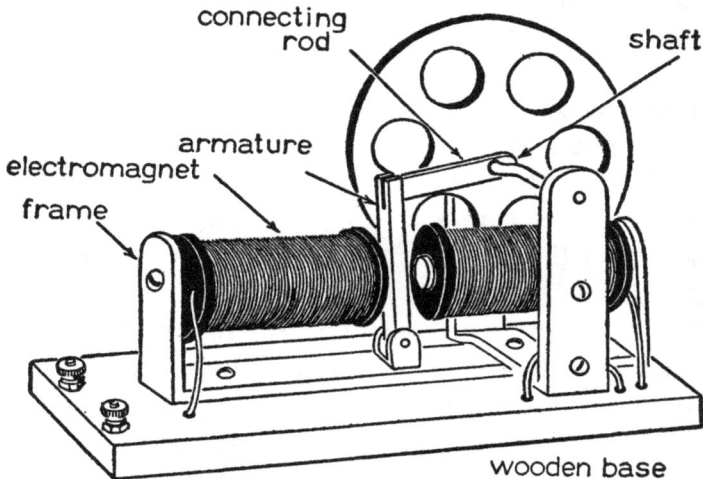

FIG. 238. — The Electric Engine.

The Base is made of a piece of hardwood, seven inches long, three and one-half inches wide and one-half an inch thick.

The Magnets. The size of the electromagnets will largely determine the dimensions of the rest of the engine. The magnets are made of three-eighths inch round iron, two and one-half inches long, provided with two fibre washers one and one-eighth inches in diameter. One end of each of the steel cores is drilled and tapped to receive an $\frac{8}{32}$ iron screw. The experimenter may possibly be able to secure some old magnet cores fitted with fibre heads from an old telephone bell or "ringer" as they are sometimes called. A suitable bolt can be made to serve the purpose by cutting it off to the right dimensions with a hack saw. If a drill and a tap are not available for drilling and tapping the end so that the core can be properly mounted in the frame of the engine it is possible by the exercise of a little ingenuity to use the threaded portion of a bolt to good advantage. The hole in the frame should then be made larger so that the end of the bolt will slip through instead of an $\frac{8}{32}$ screw and the core clamped in position by a nut on each side.

The fibre washers are spaced two and one-sixteenth inches apart. The space in between should be wound full of No. 18 B. & S. Gauge cotton covered magnet wire. Before winding in the wire, cover the core with a layer of paper so that the wire does not touch the metal. The ends of the wire should be led out through small holes in the fibre heads.

It is not absolutely necessary to use No. 18 B. & S. Gauge wire in winding the magnets but this is the size which will give the best results on the average battery. If you use larger wire, the engine will require more current from the battery. If you use finer wire, a battery of a higher voltage will be necessary. The current consumption will, however, be less.

The Frame. The electromagnets are mounted in the frame of the engine by means of two screws passing through the holes E and D. The frame is made of a strip of wrought iron or cold rolled steel, nine and one-quarter inches long, an inch wide and one-eighth of an inch thick. The ends of the strip are rounded and bent at right angles so as to form a U-shaped piece

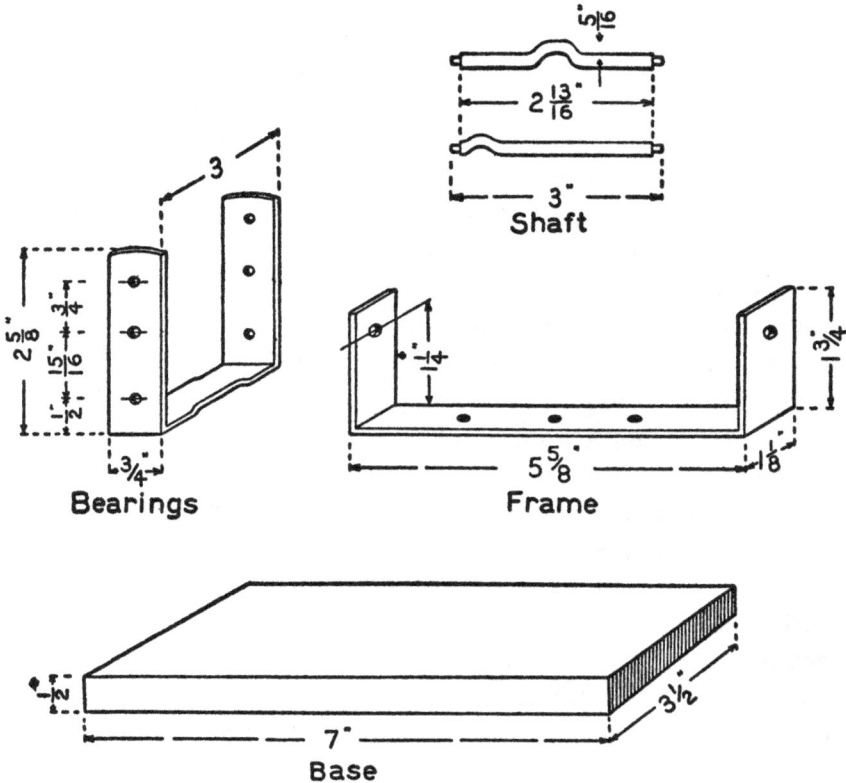

FIG. 239. — Engine Parts.

with sides one and three-quarters inches high. The holes D and E should be large enough to pass an $\frac{8}{32}$ screw. The holes A, B and C should be about one-eighth of an inch in diameter. They are used to pass the screws which hold the frame of the engine to the wooden base.

The Bearings are U shaped and are made out of a strip of iron or brass in the same manner as the frame of the engine but are three-quarters of an inch wide instead of an inch and one-eighth. The dimensions will be understood best by referring to the drawing. The $\frac{3}{32}$ inch holes near the top of each side are the bearing holes for the ends of the shaft. The one-eighth inch holes just below are used to fasten the brush holder in position. The holes in the bottom serve to fasten the bearings to the base.

The Shaft will probably prove the most difficult part of the engine to make properly. It is made of a piece of one-eighth inch steel rod bent so that it has a "throw" of one-half an inch, that is, is offset one-quarter of an inch so that the connecting rod moves back and forth a distance of one-half an inch. The finished shaft should be three inches long. The piece of steel used should be longer than this, so that it can be cut off to exact dimensions after the shaft is finished. A second crank should be bent in one end of the shaft so as to form an offset contact for the brushes. This second crank will have to be at right angles to the first one and should be much smaller. The ends of the shaft are turned or filed down to a diameter of three-thirty seconds of an inch for a distance of about the same amount so that they will fit in the bearing holes and turn freely but not allow the whole shaft to slip through. The work of making the shaft will require a small vise, a light hammer, files and a couple pairs of pliers. One pair of pliers should be of the round nosed type and the other a pair of ordinary square jawed side cutters. It may require two or three attempts before a perfect shaft is secured. When finished, it should be perfectly true and turn freely in the bearings. The bearings can be adjusted slightly by bending so that the shaft will fit in the holes and be free but yet not be loose enough to slip out. The shaft must turn freely.

The Armature is a strip of soft iron, two and one-eighth inches long, seven-sixteenths of an inch wide and three-sixteenths of an inch thick. A one-sixteenth inch slot, three-eighths of an inch long is cut in one end. A one-sixteenth inch hole is

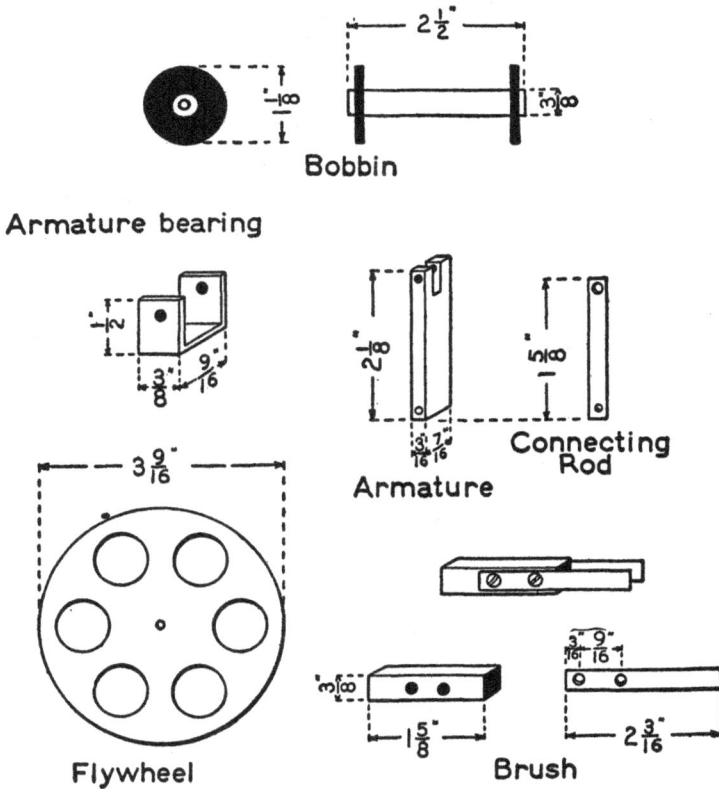

FIG. 240. — More Engine Parts.

drilled through from one side to the other, one-eighth of an inch from each end. The hole which passes through the slot is used to pass the pin which pivots the armature to the connecting rod. The other hole is used to mount the armature in its bearing. The armature bearing is a small edition of the one which is used to support the engine shaft. The armature is shown in the centre of the same illustration. The connecting rod is illustrated at the right. This is made from a strip of three-sixty-fourths inch brass, three-sixteenths of an inch wide and one and five-eighths inches long. A one-eighth inch hole should be drilled close to one end and a one-sixteenth inch hole close to the other.

The Brushes are two strips of thin phosphor bronze sheet, two and three-sixteenths inches long and nine-thirty-seconds of an inch wide. The block upon which they are mounted is hard fibre. It is one and five-eighths inches long and three-eighths of an inch square.

The Flywheel. It may be possible to secure a flywheel for the engine from some old toy. It should be about three and one-half inches in diameter. A flywheel can be made out of sheet iron or steel. It is given the appearance of having spokes by cutting six three-quarter inch holes through the face as shown. The hole in the centre of the wheel should be one-eighth of an inch in diameter. The wheel is slipped over the shaft and fastened in position by soldering.

The parts are now all ready to assemble into the complete engine. Mount the electromagnets in the frame and fasten the frame down to the wooden base so that one end of the frame comes practically flush with the left hand edge of the base. Fasten the bearing across the frame at right angles by a screw passing through the centre hole in the bottom of the bearing, through the hole A and into the base. The bottom of the bearing should be bent slightly so as to straddle the frame. The bearings should be secured and prevented from turning or twisting by two screws passed through the other two holes in the bottom. Use round-headed wood screws in mounting the bearing and the frame. The armature bearing should be mounted on the frame

directly between the two electromagnets. Then place the armature in position by slipping a piece of one-sixteenth inch brass
rod through the bearing holes and the hole in the lower part of
the armature.

Solder the flywheel in position on the shaft and snap the latter
into the bearings. Adjust the bearings so that the shaft will turn
freely. The connecting rod should be slipped over the shaft before it is placed in the bearings. Fasten the other end of the
connecting rod to the armature by means of a piece of one-sixteenth inch brass rod which passes through the small holes bored
for that purpose. When the flywheel is spun with the fingers, the
armature should move back and forth between the two electromagnets and almost but not quite touch the two magnet poles.
All the moving parts should be fitted firmly together but be free
enough so that there is no unnecessary friction and so that the
engine will continue to run for a few seconds when the flywheel
is spun with the fingers.

The brushes, supported on their fibre block, should be
mounted on the bearing by means of two screws passing
through the holes in the bearing into the block. The position of
the brushes should be such that the shaft passes between the two
upper ends but does not touch them unless the small contact
crank mentioned above is in proper position to do so. The adjustment of the brushes, so that they make contact with the
shaft at the proper moment, will largely determine the speed
and power which the finished engine will develop.

Two binding posts should be mounted on the right hand end
of the base so that the engine can be easily connected to a battery. Connect one terminal of the right hand electromagnet to
one of the binding posts. Run the other terminal of the electro-
magnet to the brush on the opposite side of the shaft. Connect
one terminal of the left hand electromagnet to the other binding

post and run the other terminal to the brush on the opposite side of the shaft. Save for a few minor adjustments, the engine is now ready to run. Connect two or three cells of dry battery to the two binding posts and turn the flywheel so that it moves from right to left across the top. Just as the crank passes "dead centre" and the armature starts to move back away from the left hand magnet, the small contact crank should touch the left hand brush and send the current through the right hand magnet. This will draw the armature over to the right. Just before the armature gets all the way over to the right, the contact should break connection with the left hand brush and interrupt the current so that the inertia of the flywheel will cause it to keep moving and the armature to start to move over towards the left hand magnet at which point the contact on the shaft should commence to bear against the right hand brush, thus throwing the left hand magnet into circuit and drawing the armature over to that side. If the brushes and the cranks are in proper relation to each other the engine will continue to repeat this operation and gradually gain speed until it is running at a good rate.

The appearance of the engine can be improved by painting the metal parts.

Chapter XVIII

DYNAMOS

THERE is perhaps no other electrical device entering into the young experimenter's domain requiring the careful workmanship and tool facilities that the dynamo does. In order to construct a practical working dynamo it is necessary to have at hand a lathe for turning the castings.

It is possible to alter an old telephone magneto so that it may be made to serve as a small dynamo. Telephone magnetos, also sometimes called hand generators, are used in some telephone systems to supply the current which rings the telephone bell at the other end. The magneto is placed in a small box on the telephone, only the handle being exposed. In order to make a call the handle is given several brisk turns before raising the receiver. When the handle is turned the moving parts of the generator revolve and produce a current of electricity which goes forth over the line and rings the bell at the other end.

Telephone magnetos have been discarded by all the large telephone systems, a method known as "central energy," in which the current for ringing bells is supplied from the central office, taking their place. There are telephone magnetos to be found in second-hand shops and at electrical houses, where they can be purchased for a fractional part of the original cost. Fifty cents will often buy a first-class second-hand telephone magneto.

Before constructing any form of dynamo a careful study of the principles of the dynamo is well worth the time spent.

Almost any book on physics or electricity, or even the encyclopedia, will be found to contain a description of the wonderful machine that supplies the power for running the trolley cars, electric lights, etc.

It will be remembered that if a bar magnet is suddenly plunged into a hollow coil of wire, a momentary electric cur-

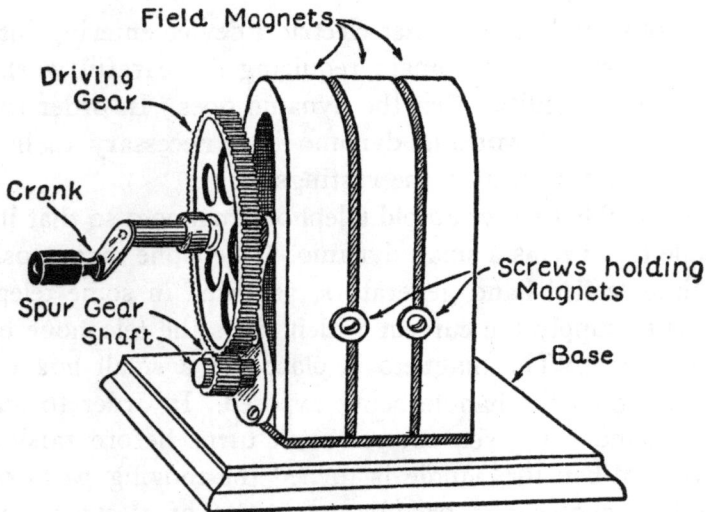

Field Magnets

Driving Gear

Crank

Spur Gear
Shaft

Screws holding Magnets

Base

FIG. 241. — A Telephone Magneto. It is wound to deliver about 50 volts A.C.

rent will be generated in the coil. The current is easily detected by means of an instrument called a galvanometer. The space in the vicinity of a magnet is filled with a peculiar invisible force called magnetism. The magnetism flows along a certain path, passing through the magnet itself and then spreading out in curved lines. If a sheet of paper is laid over a magnet and a few iron filings are sprinkled on the paper, they will follow the magnetic lines of force.

When the magnet is plunged into the hollow coil, the lines

of force flow through the turns of wire, or are said to cut them. Whenever lines of force cut a coil of wire and they are in motion, electricity is produced. It does not matter whether the coil is slipped over the magnet or the magnet is plunged into the coil, a current will be produced as long as they are in motion. As soon as the magnet or the coil stops moving the current stops.

By arranging a coil of wire between the poles of a horseshoe magnet so that it can be made to revolve, the motion can be made continuous and the current of electricity maintained.

Some means of connection with the coil of wire must be established so that the current can be led off. If two metal rings are connected to the ends of the coil, connection can be made by little strips of metal called brushes rubbing against the rings. This scheme is the principle of the telephone magneto and the basis of all dynamos.

In the telephone magneto, more than one horseshoe magnet is usually provided. The coil of wire revolves between the poles

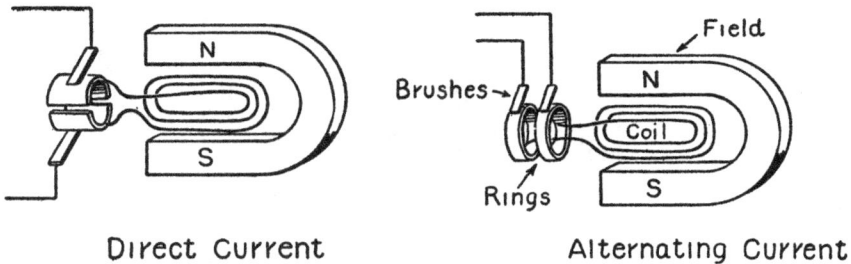

Direct Current   Alternating Current

FIG. 242. — The Principle of the Alternator and the Direct-Current Dynamo.

of the magnets. The coil is wound around an iron frame and together they are called the armature. The end of the armature shaft is fitted with a small spur gear meshing with a larger gear bearing a crank, so that when the crank is turned the motion is multiplied and the armature is caused to revolve rapidly. One end of the coil or armature winding is connected to a small brass pin. This pin connects with a second pin set in the end of the shaft in an insulating brush of hard rubber. The other terminal of the coil is connected to the armature itself. Thus connection can be had to the coil by connecting a wire to the frame of the machine and to the insulated pin. Contact with the pin is established by a brush or spring.

The armature of a magneto is usually wound with a very fine silk insulated wire, about No. 36 B. & S. gauge in size. This

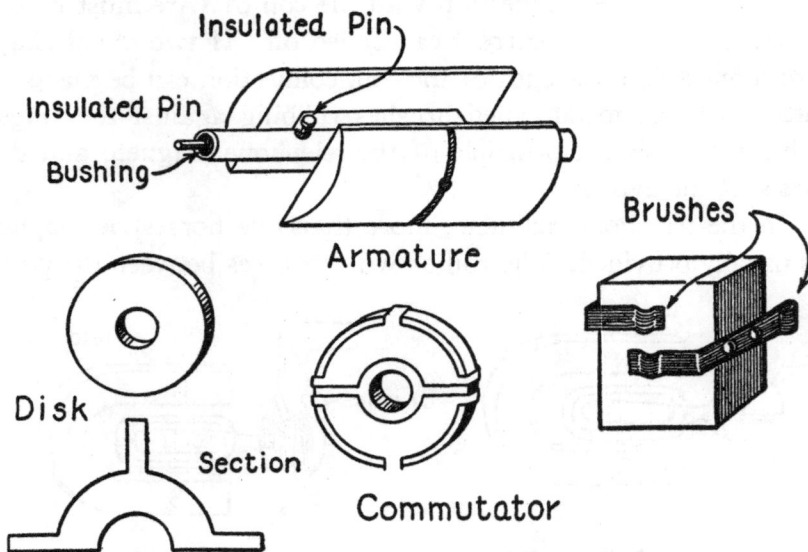

FIG. 243. — Details of the Armature, Commutator, and Brushes for changing the Magneto into a Dynamo and Motor.

should be carefully removed and wound upon a spool for future use. Replace the wire with some ordinary cotton-covered magnet wire, about No. 24 or 25 B. & S. gauge, winding it on very carefully and smoothly. Connect one end of the winding to the pin leading to the insulated pin by soldering it. This pin is the one at the end of the shaft opposite to that one to which the spur gear is fastened. Connect the other end of the wire to the pin at the same end of the shaft as the gear. This pin is grounded, that is, connected to the frame.

An ordinary telephone magneto gives a very high voltage current. The voltage may vary from twenty-five to several hundred, depending upon how fast the machine is run. This is due to the fact that the armature winding is composed of a very large number of turns of wire. The more turns that are placed on the armature, the higher its voltage will be. The current or amperage of a large telephone magneto wound with a large number of turns of fine wire is very low. Too low, in fact, to be used for anything except ringing a bell or testing. Winding the armature with fewer turns of large wire reduces the voltage and increases the amperage so that the current will light a small lamp or may be used for other purposes. The winding does not change the principle of the magneto, it merely changes its amperage and voltage.

The magneto may be mounted on a wooden base-board and screwed to a table, so that the handle may be turned without inconvenience. A small strip of copper, called a brush, should be fastened to the base with a screw and brought to bear against the end of the insulated pin. The brush should be connected to a binding-post with a piece of wire. A second wire leading to a binding-post should be connected to the frame of the magneto. When the handle is turned rapidly, currents may be drawn from the two binding-posts.

The current is of the kind known as alternating, that is to say, it flows first in one direction, then reverses and flows in the other. It will light a small incandescent lamp, ring a bell or operate a small motor.

In order to make the machine give direct current, it must be fitted with a commutator. This is somewhat difficult with some magnetos but the following plan may be carried out in most cases. Cut a small fiber disk about one inch in diameter from sheet fiber three-sixteenths of an inch thick. Cut a small hole in the center, just large enough so that the fiber will slip very

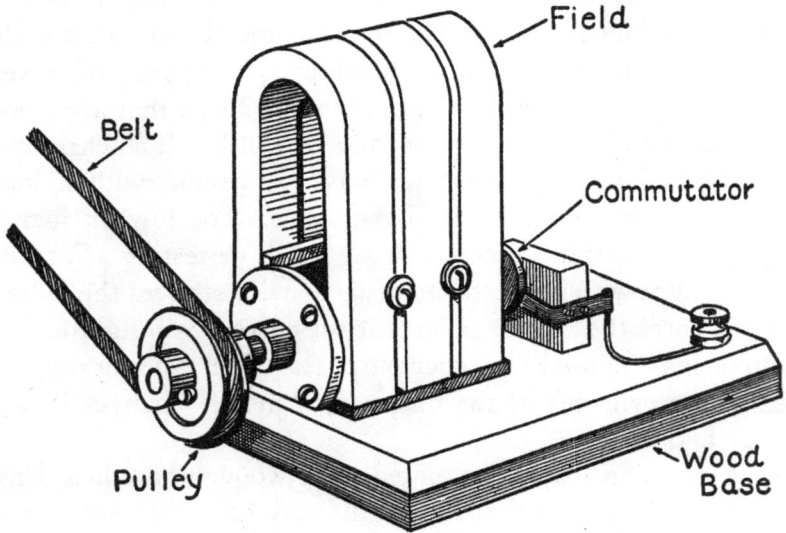

FIG. 244. — The completed Generator and Motor.

lightly over the end of the shaft from which the insulated pin projects. Two small commutator sections must be cut from sheet-brass or sheet-copper. The three long ears shown in the drawing are bent back around the fiber and squeezed down flat with a pair of pincers so that they grip the fiber very tightly and will not slip. One ear on one section should be bent over the back down to the hole, where it will connect with the shaft. The other section of the commutator is connected to the insulated pin by a drop of solder. In this manner, one end of the winding is connected to one section of the commutator and the other end to the other section. The commutator should fit tightly on the end of the shaft so that it will not twist. The dividing line between the section should be parallel to a line drawn to the axis of the actual armature coil. When the iron parts of the armature are nearest the poles of the horseshoe magnets in their revolution, the slot in the commutator should be horizontal.

When the magnet is provided with a commutator, it may also be run as a motor by connecting it to a battery. In order to operate it either as a dynamo or a motor, however, it must first be fitted with a pair of brushes. They are made from two small strips of sheet-copper bent as shown and mounted on a small wooden block. They must be adjusted to bear against the commutator so that when the dividing line between the two sections is horizontal, the upper brush bears against the upper section and the lower brush against the lower section. The two brushes form the terminals of the machine. They should be connected to binding-posts.

In order to operate the dynamo properly and obtain sufficient current from it to operate a couple of small incandescent lamps, it will have to be provided with a pulley mounted on the end of the shaft after the gear wheel has been removed. The dynamo may then be driven at high speed by connecting it to a sewing-machine with a belt, or the back wheel of a bicycle from which the tire has been removed. A heavy cord may be used as a belt.

The voltage and amperage of the dynamo will depend upon the machine in question, not only upon the size of the wire but also upon the size of the machine, the speed at which it is run, and the strength of the horseshoe magnets. It is impossible to tell just what the current will be until it is tested and tried.

A 10-WATT DYNAMO

Probably few experimenters fully understand how almost impossible it is to construct a dynamo, worthy of the name as such, without resort to materials and methods employed in the commercial manufacture of such machines. Practical telegraph instruments, telephones, etc., can be constructed out of all sorts of odds and ends, but in order to make a real dynamo it is necessary to use certain materials for which nothing can be substituted.

Except in special instances, the *field magnets* must be soft gray cast-iron.

The wire used throughout must be of good quality and must be new.

The necessity for good workmanship in even the smallest detail cannot be overestimated. Poor workmanship always results in inefficient working. No dynamo will give its stated output continuously and safely unless the materials and workmanship are up to a high standard.

Since castings must be used as field magnets, a pattern is necessary to form the mould for the casting. Pattern work is something requiring skill and knowledge usually beyond the average experimenter. A lathe is necessary in order to bore or tunnel the space between the ends of the field magnet into which the armature fits.

It may be possible for several boys to club together and have a pattern made by a pattern-maker for building a dynamo. Then by using the lathe in some convenient shop or manual training school secure a field magnet and armature for a really practical small dynamo.

For these reasons, I have described below a small dynamo of about ten watts output, the castings for which can be purchased from many electrical dealers with all machine work done at an extremely low price.

The field magnet shown in Figure 245 is drawn to scale and represents the best proportions for a small "overtype" dynamo of ten to fifteen watts output.

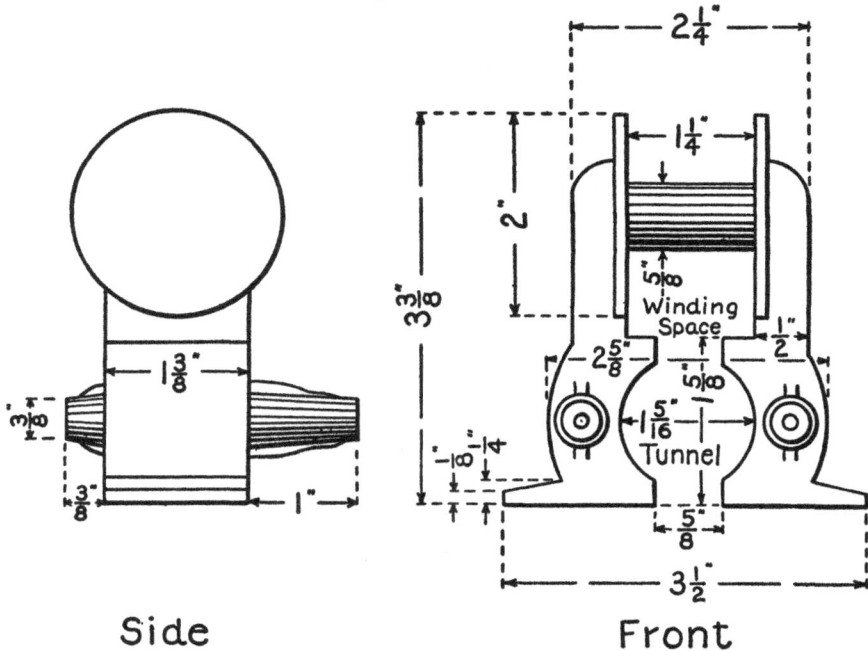

Side   Front

FIG. 245. — Details of the Field Casting.

The dimensions are so clearly shown by the drawings that further comment in that respect is unnecessary.

The armature is of the type known as the "Siemen's H." It is the simplest type of armature it is possible to make, which is a feature of prime importance to the beginner at dynamo construction, although it is not the most efficient form from the electrical standpoint. The armature in this case is also a casting and therefore a pattern is required.

The patterns for both the field and the armature are of the same size and shape as shown in Figures 245 and 246. They are made of wood and are finished by rubbing with fine sand-paper until perfectly smooth and then given a coat of shellac. The parts are also given a slight "draft," that is, a taper toward

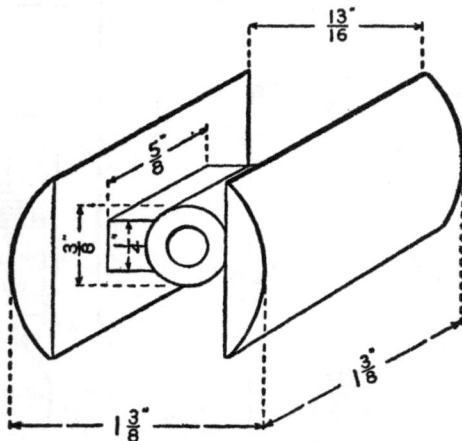

FIG. 246. — Details of the Armature.

one side, so that the pattern may be withdrawn from the mould at the foundry.

The patterns are turned over to a foundry, where they are carefully packed in a box, called a "flask," full of moulder's sand. When the patterns are withdrawn, they will leave a per-fect impression of themselves behind in the sand. The mould is then closed up and poured full of molten iron. When the iron has cooled the castings are finished except for cleaning and boring.

The shaft is a piece of steel rod, three-sixteenths of an inch in diameter, and four and one-half inches in length.

The portion of the field into which the armature fits is bored out to a diameter of one and five-sixteenth inches. Considerable care is necessary in performing this operation in order not to break the field magnet apart by taking too heavy a cut.

The armature should be turned down to a diameter of one and one-quarter inches or one-sixteenth of an inch smaller than the tunnel in which it revolves between the field magnets. The center of the armature is bored out to fit the shaft.

Figure 247 shows a two-part commutator for fitting to an armature of the "Siemen's H" type. It consists of a short piece of brass tubing fitted on a fiber core and split lengthwise on two opposite sides, so that each part is insulated from the other.

The fiber is drilled with a hole to fit tightly on the shaft. It is then placed in a lathe and turned down until a suitable piece of brass tube can be driven on easily.

Two lines are then marked along the tube diametrically opposite. A short distance away from each of these lines, and on each side of them, bore two small holes to receive very small wood screws. The screws should be countersunk. It is very important that none of the screws should go into the fiber core far enough to touch the shaft.

The commutator may then be split along each of the lines between the screws with a hacksaw. The saw-cut should be continued right through the brass and slightly into the insulating core. The space between the sections of the commutator should be fitted with well-fitting slips of fiber, glued in.

The commutator should now be trued up and made perfectly smooth.

The commutator is provided with a small brass machine screw threaded into each section near the edge as shown in Figure 247. These screws are to receive the ends of the armature winding and so facilitate connections.

The commutator, shaft and armature are assembled as shown in Figure 247.

The armature may be held to the shaft by a small set screw or a pin. The commutator should fit on the shaft very tightly so that it will not slip or twist.

Every part of the armature and shaft touched by the armature winding must be insulated with paper which has been soaked in shellac until soft. The armature must be left to dry before winding.

The armature should next be wound with No. 20 B. & S. gauge single-cotton-covered magnet wire. Sufficient wire

FIG. 247. — The Commutator. The illustration shows how to connect the Armature Winding to the Commutator.

should be put on to fill up the winding space completely. Care should be taken, however, not to put on too much wire or it will interfere with the field magnets and the armature cannot revolve. After winding the armature, test it carefully to see that the wire is thoroughly insulated from the iron.

If the insulation is correct, paint the whole armature with thick shellac varnish and bake it in a warm oven to set the shellac. Do not close the door of the oven or the temperature may become too high and char the shellac.

Figure 247 is a diagram showing how the winding should be wound about the armature, always in the same direction, just as if the armature were an ordinary electro-magnet.

The ends of the winding are each connected to one of the

commutator sections by scraping the wire and placing it under the screws.

The winding space in the field magnet should be shellacked, and insulated with brown paper by wrapping the core with a strip of paper and covering the bobbin ends with circular pieces made in two halves.

The field magnet is wound full of No. 20 B. & S. gauge single-cotton-covered wire. The wire should be put on in smooth, even layers and the winding space completely filled.

The base for the dynamo is a piece of hard wood, five inches long, four inches wide and five-eighths of an inch thick.

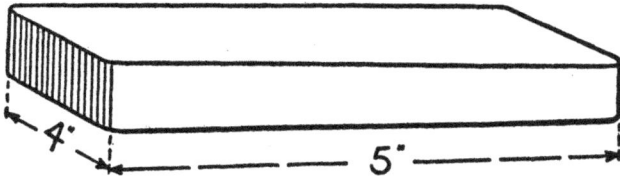

FIG. 248. — The Base for the Dynamo is a piece of Hardwood.

The bearings are small brass castings. It is necessary first to make a wooden pattern and send it to the foundry for the castings.

The bearings are fastened to the projecting arms on the field casting by means of $\frac{8}{32}$ machine screws.

The field magnet should not be screwed down on to the base until the armature runs easily and truly in the tunnel.

The brushes are made from thin gauge sheet-copper.

They are bent at right angles and mounted on the base on either side of the commutator with small round-headed wood screws.

One end of the dynamo shaft is provided with a small pulley to accommodate a small leather belt.

The dynamo is connected as a "shunt" machine, that is, one terminal of the field magnet is connected to one of the brushes, and the other terminal to the other brush.

A wire is then led from each of the brushes to a binding-post.

A shunt dynamo will only generate when run in a certain direction. In order to make it generate when run in the op-

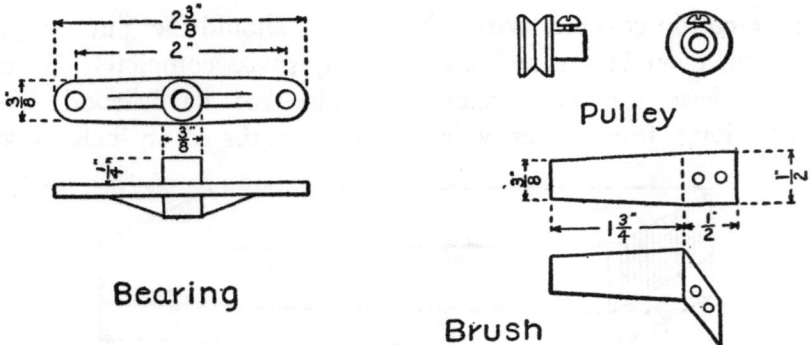

FIG. 249. — The Bearings are brass Castings. The Pulley is turned. The brushes are cut out of Sheet metal.

posite direction, it is necessary to reverse the field connections.

The dynamo just described should have an output of from 10 to 15 watts and deliver about 6 volts and 1¾ to 2½ amperes.

In order to secure current from the dynamo it will first be necessary to magnetize the field by connecting it to several batteries.

It will be found that the dynamo will also operate as a very efficient little motor, but that on account of having a two-pole armature it must be started by giving the shaft a twist.

It can be used as a generator for lighting small lamps, electro-plating, etc., but cannot be used for recharging storage cells on account of having a two-pole armature.

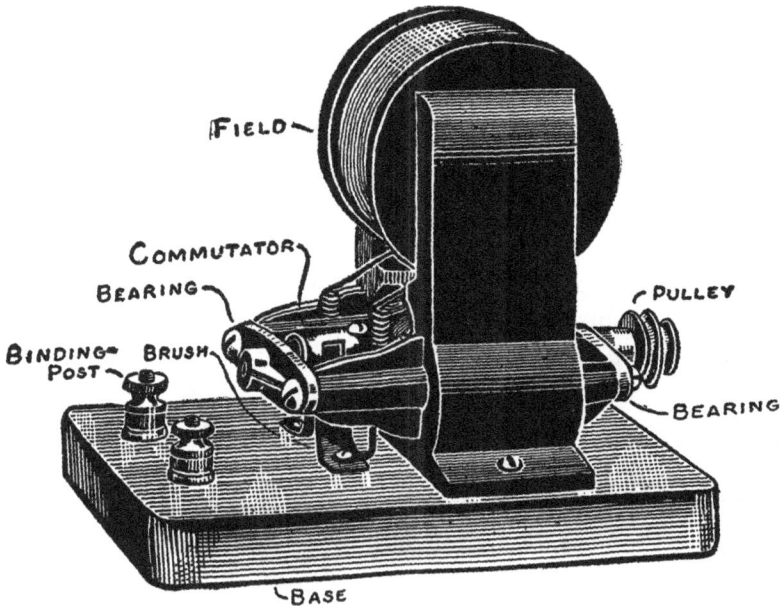

FIG. 250. — Complete 10-watt Dynamo.

The dynamo may be driven with a small water motor or from the driving-wheel of a sewing-machine.

Before the machine will generate as a dynamo, it must be connected to a battery and run as a motor. This will give the field the "residual magnetism" which is necessary before it can produce current itself.

CHAPTER XIX

AN ELECTRIC RAILWAY

No toys loom up before the mind of the average boy with more appeal to his love of adventure than do railway cars and trains. The construction and operation of miniature railways is the hobby not only of boys but of grown men.

The height of ambition of many boys is not only to own a miniature railway system but to build one. The car shown in the illustration is a model of the trolley car of thirty or forty years ago. Such cars were once an important part of America's transportation system. It can be built by any boy willing to exercise the necessary care and patience in its construction.

The first operation is to cut out the floor of the car. This is a rectangular piece of hard wood, eight inches long, three and one-quarter inches wide and one-half of an inch thick. Its exact shape and dimensions are shown in the sketch.

The rectangular hole cut in the floor permits the belt which drives the wheels to pass down from the countershaft to the axle.

The two pieces forming the wheel-bearings are cut out of sheet brass according to the shape and dimensions shown. The brass should be one-sixteenth of an inch thick. The two projecting pieces at the top are bent over at right angles so that they can be mounted on the under side of the car floor by small screws passing through the holes. The holes which form the bearings for the ends of the axles upon which the

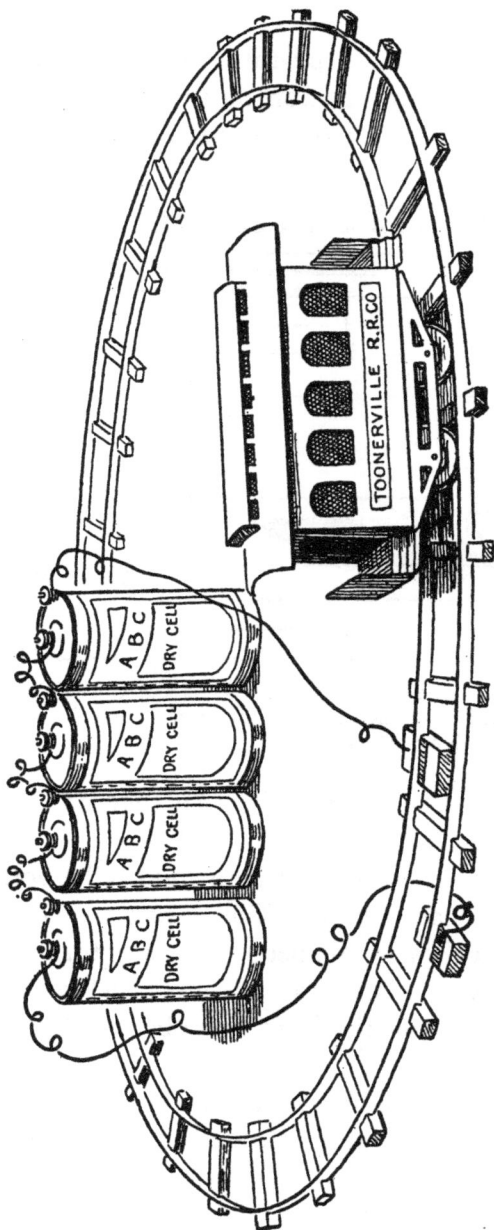

Fig. 251. — Complete Electric Railway Operated by Dry Cells.

wheels are mounted should be three inches apart. The bearings cannot be placed in position on the under side of the car floor until the wheels and axles are ready, but when this work

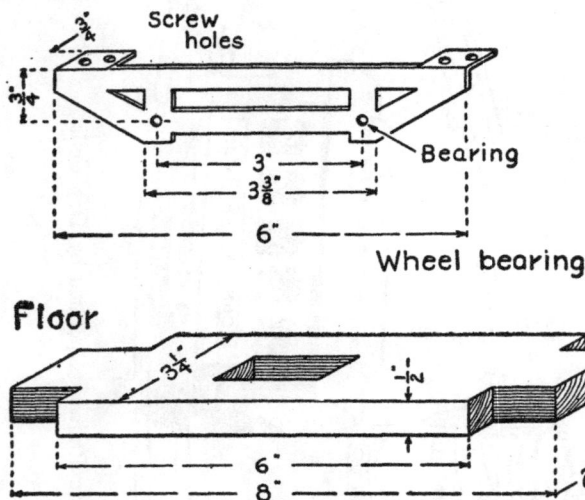

FIG. 252. — The Floor of the Car is cut out of wood. The Bearings which support the Wheels and Axles are made of Sheet Metal.

is done, care should be taken to see that they line up and come exactly opposite each other.

The wheels themselves cannot be made by the young experimenter unless he has a lathe. They are flanged wheels, one and one-eighth inches in diameter, and are turned from cast iron or brass. Such wheels can be purchased ready made, or it may be possible to obtain from some broken toy a set which will prove suitable.

Each shaft is composed of two pieces of "Bessemer" rod held together by a short piece of fiber rod having a hole in each end into which one end of each piece of iron rod is driven. The wheels fit tightly on the other end of each of these pieces.

They should be spaced so as to run on rails two inches apart.

The purpose of the fiber rod is to insulate the halves of the axle from each other. The electric current which operates the car is carried by the two rails which form the track, and if the axles were made in one piece or the halves joined together so as to form an electrical connection, the battery furnishing the current would be short-circuited, because the current would pass along the two rails and across the axles instead of through the motor.

One pair of wheels are fitted with a grooved pulley one inch in diameter.

It is hardly necessary to say that the wheels and axles should be perfectly aligned, and should run true.

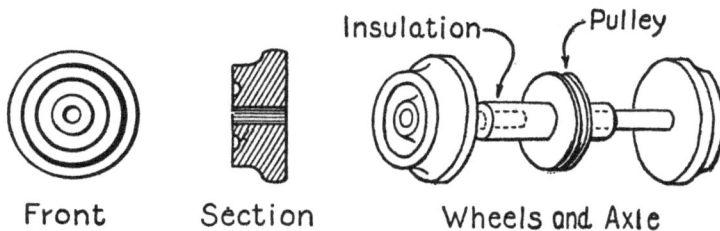

Front   Section   Wheels and Axle

FIG. 253. — The Wheels and Axle.

The motor used to drive the car will prove more satisfactory if purchased ready made. A self-starting three-pole motor similar to that shown in Figure 254 will serve very nicely. The wooden base should be removed and the motor screwed down firmly to the floor of the car.

One terminal of the motor is connected to one of the bearings, and the other terminal to the other bearing.

The motor is belted to a countershaft so that it will have sufficient power to move the car. It cannot be directly connected or belted to the axle, because the speed of a small motor is so high that it has comparatively little turning power or *torque*. The speed must be reduced and the torque increased before it will drive the car.

The countershaft consists of two grooved pulleys mounted upon an axle running in two bearings mounted upon the floor of the car. The bearings are made from a strip of heavy sheet-brass, bent at right angles and fastened to the car floor with

FIG. 254. — A self-starting three pole Motor is used to drive the Car.

small screws. The large pulley is one inch and one half in diameter and the small pulley is five-sixteenths of an inch in diameter. The countershaft is mounted in such a position that a belt may be run from the small pulley to the pulley mounted on the axle of one pair of wheels. A belt is also run from the small pulley on the motor to the large pulley on the countershaft. The pulleys must all be carefully lined up so that the belts will run in their grooves without danger of slipping out.

The shield on the platform at each end of the car is made of sheet-iron or tin. Two small projections on the bottom are bent over at right angles and used to secure the shields in position by driving a small tack through them into the floor of the car.

The steps on either side of each platform are also made by bending strips of sheet-iron or tin and fastening them to the car with small nails or tacks.

The coupler consists of a strip of tin having a small hook

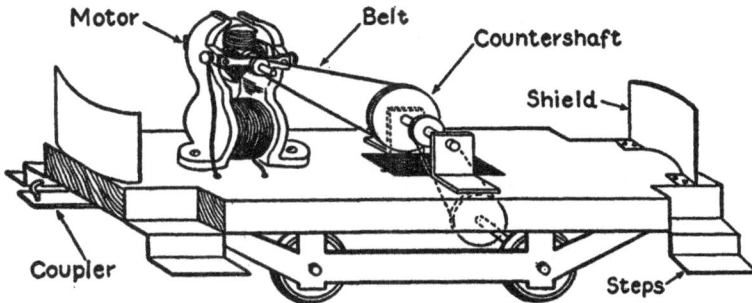

FIG. 255. — The Car Truck without the Body showing how the Motor drives the Wheels through a Countershaft.

soldered to the end so that a trail car may be attached if desirable.

The car is now ready for testing, and when held in the hand so that the wheels are free to run, two cells of dry battery should be found all that is necessary to drive them at a fair rate of speed. The two wires leading from the battery should be connected to the bearings, one wire leading to each bearing. It will require more than two cells, however, to drive the wheels properly when the car is on the track. All moving parts should run freely and smoothly. The car may be used just as it is, but if fitted with a body and a top it will present a much more realistic appearance.

The sides and ends of the car body are made of sheet-iron or tin. They may be made from one piece of metal eighteen and one-half inches long and three and three-quarters inches wide. The doors and windows are cut out with a pair of tin-snips. The small projections along the top are bent down at right angles and the roof is fastened to them. The dotted lines indicate the places for bending these projections and also the sides and ends of the car.

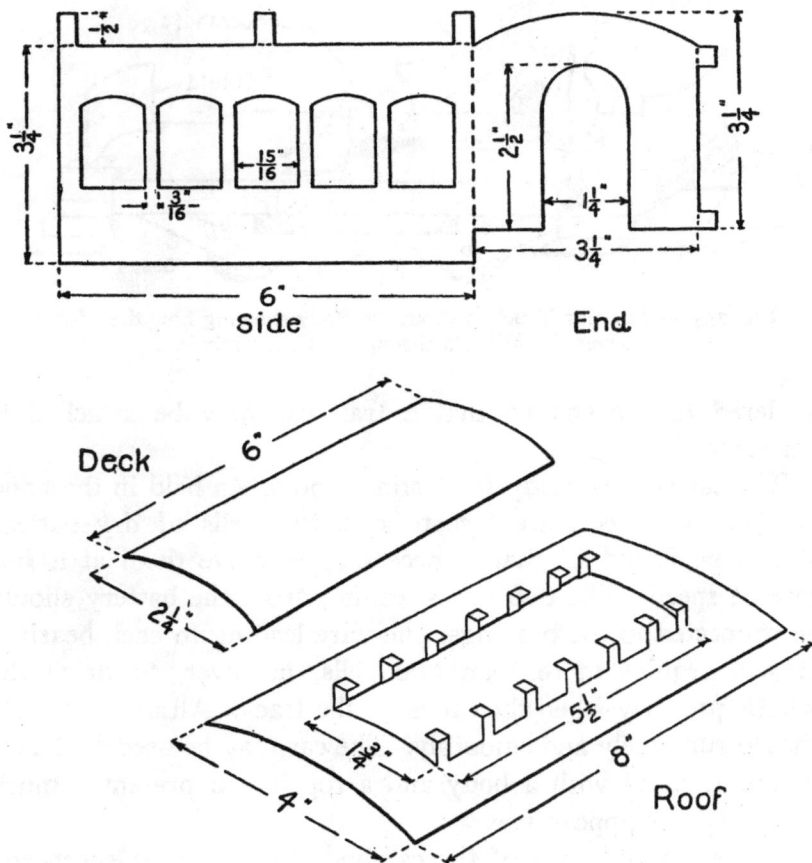

FIG. 256. — Patterns for the Body of the Car.

The roof is made in two pieces. It also is sheet-iron or tin. The roof proper is eight inches long and four inches wide. It has a hole five and one-half inches long and one and three-quarters inches wide cut in the center. A number of small pro-

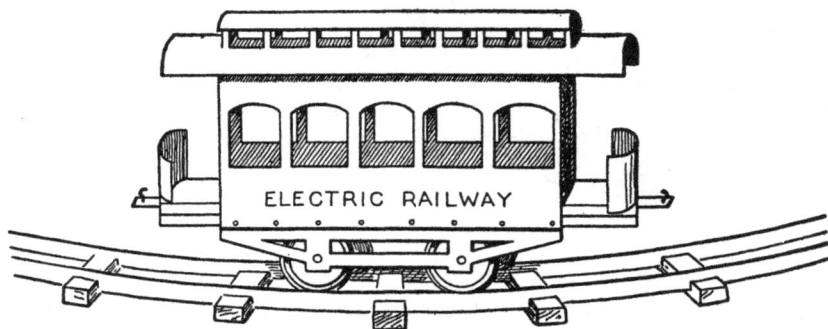

FIG. 257. — The Completed Car.

jections are left and bent upward to support the deck and to form imitation ventilators. The deck is six inches long and two and one-quarter inches in width. It is placed in position on the roof and fastened by soldering. The roof is fastened to the sides and ends of the car by soldering. It must be bent slightly to conform with the curve at the top of the front and the rear of the car.

The track is made of spring brass one-half inch wide and either No. 20 or No. 22 gauge in thickness.

The wooden ties are three and one-half inches long, three-quarters of an inch wide and three-eighths of an inch thick. Each tie has two saw-cuts, exactly two inches apart across the top face. This part of the work is best performed in a miter-box so that the cuts will be perfectly square across the ties. A saw should be used which will make a cut of such a size that the steel track will fit tightly into it.

The distance between the two rails of the track, or the "gauge," as it is called, is two inches.

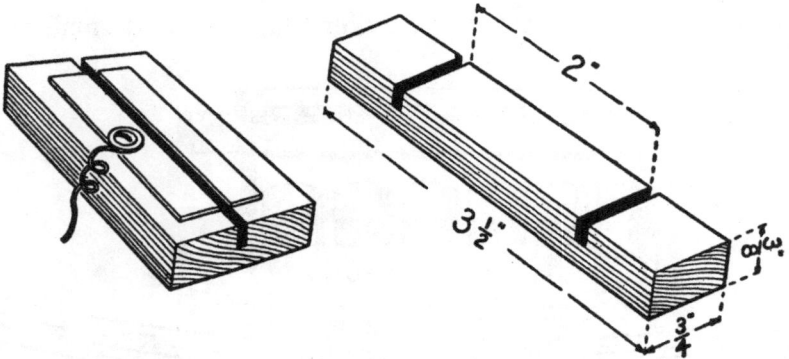

FIG. 258. — Details of a Wooden Tie and a Connector for Joining the Ends of the Rails.

The spring steel is forced into the saw-cuts in the ties by tapping with a light wooden mallet. The ties should be spaced along the track about three inches apart. The work of laying the track must be very carefully done so that the car wheels

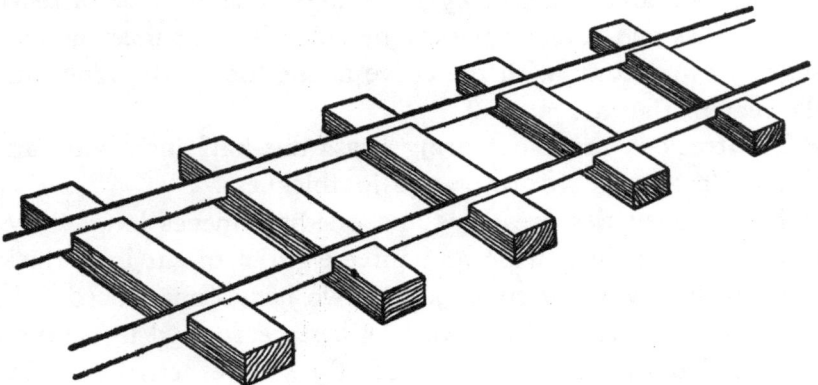

FIG. 259. — The Track is assembled by forcing the Metal Strip into the Saw cuts in the Ties.

will not bind at any spot. Curves should not be too sharp, or the car will not pass around.

The track may be laid out in a number of different shapes, some of which are shown.

A circle is the easiest form of track to make. In laying out a circle or any sort of curved track, the outside rail must necessarily be made longer than the inside one.

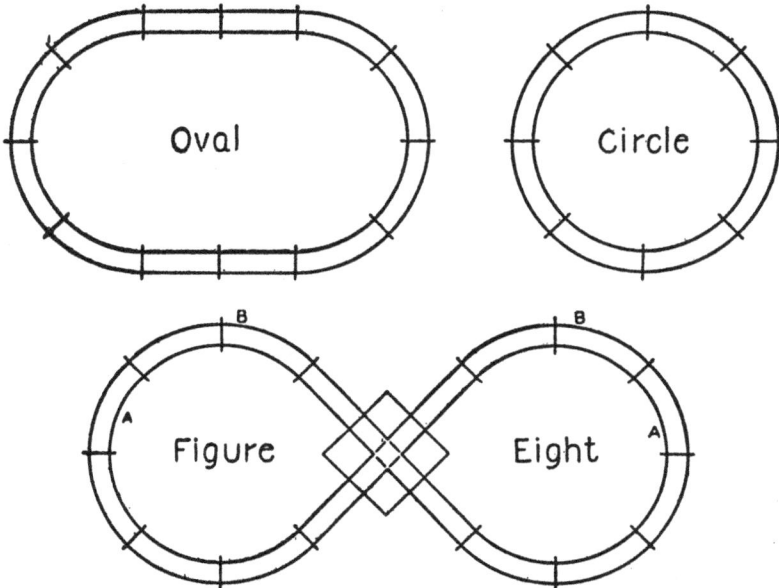

FIG. 260. — Three Different Patterns for laying out the Track.

The oval shape is a very good form to give the track in a great many cases, especially where it is desirable for the car to have a longer path than that afforded by a circle.

In order to make a figure-eight out of the track, a crossing, or "cross-over," as it is sometimes called, will be required. A cross-over permits two tracks to cross each other without interference. It consists of a wooden base, eight inches square and three-eighths of an inch thick. Four saw-cuts, each pair exactly parallel, and two inches apart, are made at right angles to each other across the top surface of the base, as shown in the illustration.

The track used on the cross-over is semi-hard hoop-brass,

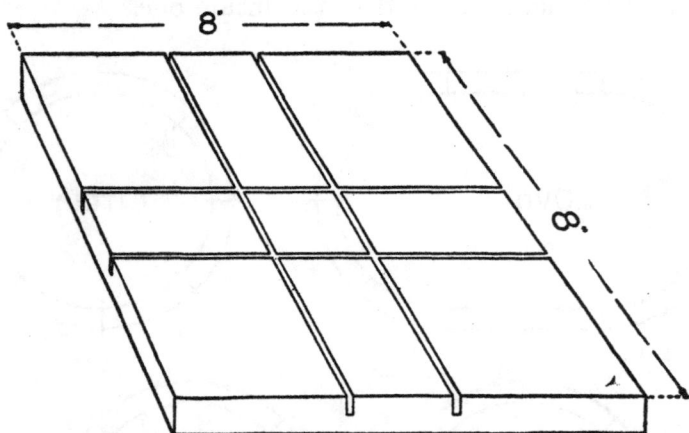

FIG. 261. — Details of the Base of the Cross-over.

one-half of an inch wide and of the same gauge as the rest of the track.

Four pieces of the brass, each five inches long, are bent at right angles exactly in the center. Four short pieces, each one and one-half inches long, will also be required.

The cross-over is assembled as shown in Figure 262. The strips marked *D* are strips of very thin sheet-brass or copper. The purpose of these strips is to connect the ends of the track on the cross-over to the ends of the track forming the figure-eight so that the cross-over will not be a "dead" section, that is, a section of track where the car cannot get any current.

The long strips, bent at right angles to each other and marked

A, A, B, B, in the illustration, are forced into the saw-cuts in the base over the strips marked *D.*

The small pieces, *C, C, C, C,* are placed in between the long strips, leaving a space between so that the flanges of the car wheels can pass. The pieces, *C, C, C, C,* should form a square open at the corners. The two long strips, *A, A,* should be at opposite corners diagonally across the square. *B* and *B* should occupy the same relative position at the other corners. *A* and

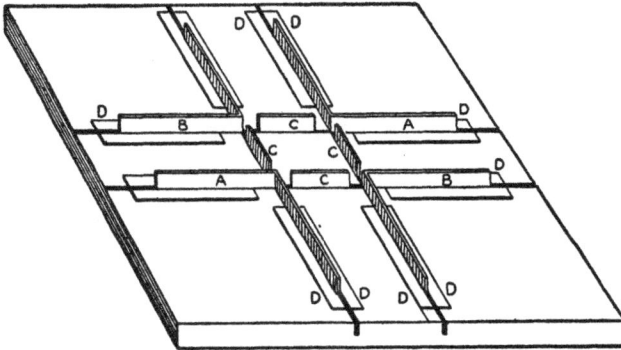

FIG. 262. — The Completed Cross-over.

A are connected together and *B* and *B* are connected together by wires passing on the under side of the base.

The ends of the track forming the figure-eight are forced into the saw-cuts at the edges of the base so that they form a good electrical connection with the small strips marked *D.*

It is necessary to use care in arranging a figure-eight track, or there will be danger of short-circuiting the batteries. The outside rails of the figure-eight, distinguished by the letter *B* in the illustration, should be connected together by the cross-over. The inside rails, marked *A,* should also be connected together by the cross-over.

In order to make a good mechanical and electrical connec-

tion between the ends of the rails when two or more sections
of track are used in laying out the system, it is necessary to
either solder the rails together or else use a connector such as
that shown in Figure 258.

This consists of a small block of wood having a saw-cut
across its upper face and a piece of thin sheet-brass set into
the cut. The two rails are placed with their ends abutting and
one of these connectors slipped up from beneath and forced on
the rails. The piece of thin brass set into the wooden block
serves to make an electrical connection between the two rails
and also to hold them firmly in position. A small screw and
a washer placed outside the track and passing through the brass
strip will allow a battery wire to be conveniently attached.

The steel rails should be occasionally wiped with machine
oil or vaseline to prevent rusting, and also to allow the car to
run more freely wherever the flanges of the wheels rub against
the rails in passing around a curve.

Four dry cells or three cells of storage battery should be
sufficient to operate the car properly. If it is desirable, a small
rheostat may be included in the battery circuit, so that the
speed of the car can be varied at will. The motor and the wheels
should be carefully oiled so that they will run without friction.
The belts should not be so tight that they cause friction or so
loose that they allow the motor to slip, but should be so adjusted that the motor runs freely and transmits its power to
the wheels.

The car may be made reversible by fitting with a small
current reverser, but unless the reverser is carefully made the
danger of loss of power through poor contacts is quite considerable. If the car is fitted with a reverser the handle should
project from the car in a convenient place where it can be easily
reached and the car sent back or forth at will.

A railway system such as this can be elaborated and extended by adding more than one car to the line and bridges and stations.

The ends of a blind section of track, that is, a straight piece of track not part of a circle or curve so that the car can re-

FIG. 263. — A Design for a Railway Bridge.

turn, should be fitted with a track bumper, to prevent the car leaving the rails.

No dimensions are given in the illustrations, showing the designs for a bridge and an old fashioned station, because

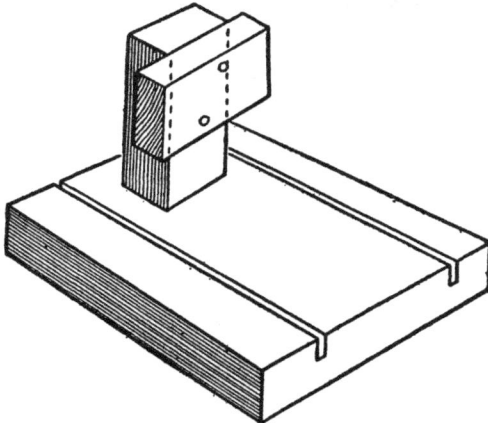

FIG. 264. — A Bumper for Preventing the Car From Leaving the Rails.

they are best determined by the scale upon which the railway system is to be extended.

The bridge is built entirely of wood, with the exception of the brass rails.

FIG. 265. — A Design for an old-fashioned Station.

The station may be made out of thin wood, such as cigar-box wood. The doors, windows, etc., may be painted on the walls. If this is carefully done, it will give a very realistic appearance to your station.

CHAPTER XX

MINIATURE LIGHTING

MINIATURE lighting is a field of many interesting possibilities for the young experimenter.

Miniature lights, operated from batteries, may be used in various ways; to light dark corners, hallways, or other places where a light is often temporarily wanted.

Miniature lighting has been made practical by the tungsten filament lamp. The filament, or wire, which becomes hot and emits the light when the current is turned on, is made of *tungsten* in a tungsten lamp. A small tungsten lamp uses only about one watt per candle-power.

Tungsten Lamps are made for voltages as low as one and one-quarter, and will light on one cell of dry battery. The range of voltages is quite wide and varied. A few of the most common sizes are given below:

MINIATURE TUNGSTEN BATTERY LAMPS

1.5 volts	for one dry cell
2.5 volts	for two-cell flashlight battery
2.8 volts	for two-cell flashlight battery
3.5 volts	for three-cell flashlight battery
3.8 volts	for three-cell flashlight battery
4 volts	4 candle-power
6 volts	2 candle-power
6 volts	4 candle-power
6 volts	6 candle-power
6 volts	8–10–12–16–20–24 candle-power

Six-volt tungsten lamps giving a light greater than six candle-power are only practical when used with storage batteries.

The filament of a tungsten lamp is much longer than that of a carbon lamp and is usually in the form of a spiral or helix.

The bases of small battery lamps, the base being the lower portion of the lamp, which is made of brass and fits into a socket or receptacle, are made in three different types: *miniature, candelabra,* and *Ediswan.*

Double contact   Single contact   Miniature

FIG. 266. — Three Types of Bases used on Miniature Lamps.

The miniature and candelabra bases have a threaded brass shell on the outside and a small brass contact-button on the bottom. They are similar except in respect to size. The miniature base is smaller than the candelabra. The Ediswan base is a plain brass shell having two pins on the side and two contacts on the bottom. This type of base is used in this country only on automobiles. The miniature and the candelabra bases are standard for battery lighting. The miniature base has many advantages over the candelabra for the young experimenter, and should be adopted in making any of the apparatus described in this chapter.

In order to form a good electrical connection between the

lamp and the power wires some sort of a receptacle or socket is necessary. The most common arrangement for this purpose is the miniature flat-base porcelain receptacle. This type of receptacle is used in places where it can be permanently fastened in position with two small screws.

FIG. 267. — A Flat-Base Porcelain Receptacle, An X'mas Tree Socket and a Pin-Socket for Miniature Base Lamps.

The Wires used to carry the current in a miniature lighting system may be of the sort known as *annunciator* or *office* wire if the wires are to be run entirely indoors. The wire should not be smaller than No. 16 B. & S. gauge. When the wires are run outdoors, on a porch, or in some other place exposed to the weather, the wire used should be rubber-covered. Hanging lights or lights intended to be adjustable should be connected with "flexible conductor." This is made of a number of very fine wires braided together and insulated with silk. The wires used in a lighting system should not in any case be longer than it is necessary to have them. When a battery is connected to a system of wires it is found that the voltage at the end of the wires is much lower than at a point near the battery. This is called voltage "drop," and is much greater as the wires grow longer. A light placed at the end of two very long wires will not burn as brightly as it would if connected to the same battery by means of short wires.

Switches can be made by following the suggestions given in Chapter VII.

The Batteries used for miniature lighting may be made up of storage cells or dry cells. Storage cells will prove the most satisfactory, provided that the experimenter has some convenient means of recharging them or of having them recharged. Storage cells will be found of especial value wherever it is desirable to operate several lights from one battery.

A home-made Leclanché cell is only suitable where one lamp is to be operated at a time. If more than one is used, the battery is liable to become polarized and the lamps will not burn brightly.

If lamps requiring more than one ampere are to be operated on dry cells, the latter should be connected in series multiple. Two sets of dry cells connected in series-multiple will give more than twice the service of a single set.

Lamps may be connected either in multiple or in series, provided that the proper voltages of both battery and lamps are used.

When they are to be connected in multiple, the voltage of the lamps should be the same as that of the battery. When they are to be used in series, the voltage of the lamps multiplied by the number used should equal the voltage of the battery. For example, suppose that you wish to use a number of six-volt lamps on a six-volt storage battery. In that case they must be connected in multiple. But if it should be that the lamps are only two-volt lamps and you wish to operate three of them on a six-volt battery you will have to place them in series.

It is sometimes desirable to arrange a lamp and two switches so that it can be turned off or on from either switch independently of the other. This is called "three-way wiring," and is a very convenient method of arranging a light in a hallway. If one switch is placed at the top of a stairway and the other switch at the bottom, a person can pass upstairs or down-

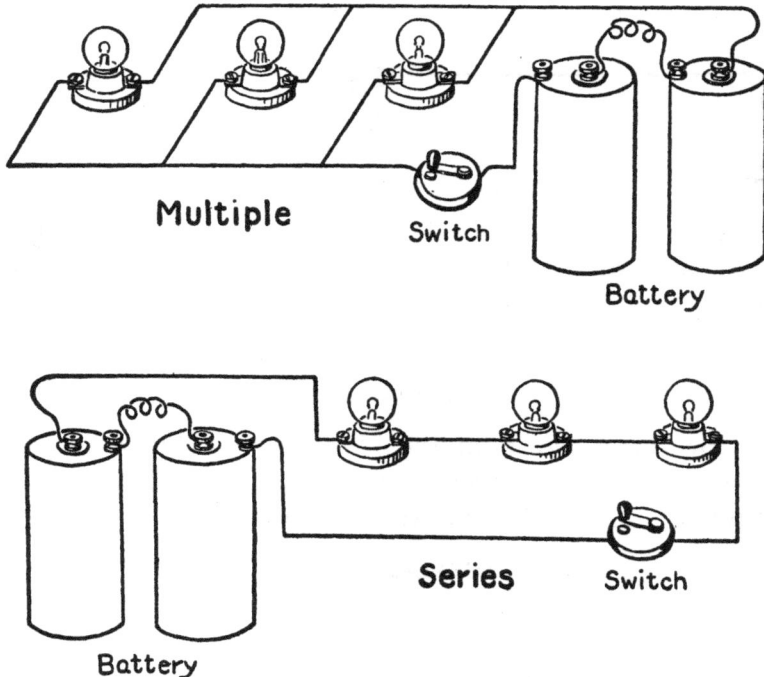

FIG. 268. — How Lamps Are Connected in Multiple and in Series.

stairs, light the lamp ahead, and turn it out as he passes the last switch, no matter in which direction the previous user of the light may have gone.

The switches are two-point switches, and the circuit should be arranged as in the accompanying diagram.

The switch-levers should always rest on one of the contacts and never be left between, as shown in the drawing. They are represented that way in the illustration in order not to conceal the contacts.

Brackets may be constructed after the plan shown in the illustration. A wooden socket or a pin-socket is mounted on the end of a small piece of brass tubing which has been bent

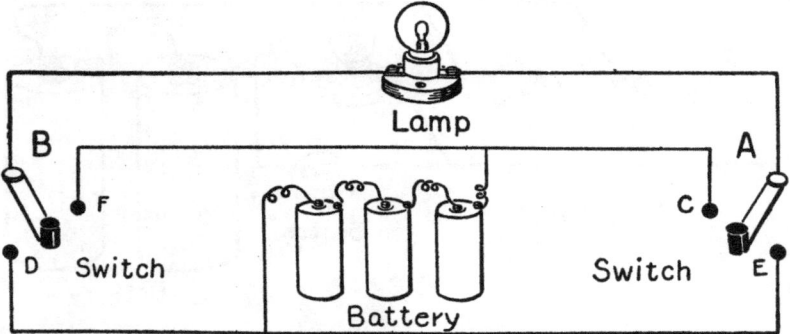

FIG. 269. — A Three-way Wiring System permits the Lamp to be turned on or off from either Switch.

into the shape shown in the illustration. The other end of the tube is set into a wooden block so that the bracket may be mounted on the wall. The wires from the socket lead through

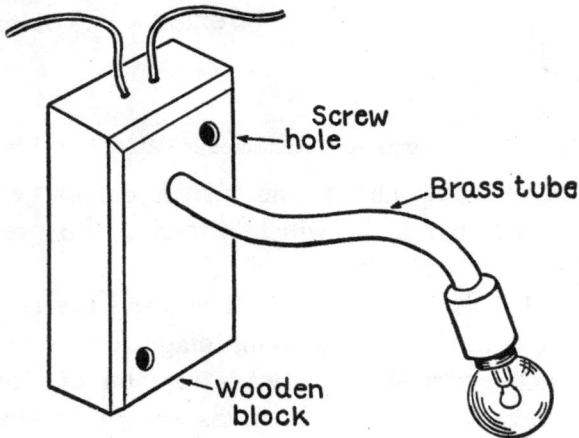

FIG. 270. — A Home-Made Bracket or Lighting Fixture.

the brass tube and through the back or top of the block.

Hanging lights may be arranged by fitting a socket and a lamp with a reflector. The reflector consists of a circular piece

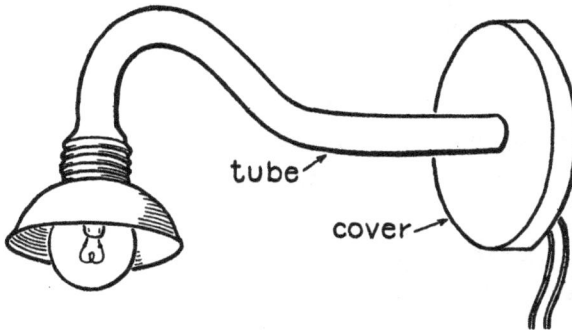

Fig. 271. — Another Type of Home-made Bracket for a Miniature Battery Lighting System.

of tin or sheet-aluminum having a hole in the center large enough to pass the base of a miniature lamp. The circle is then cut along a straight line from the circumference to the center. If the edges are pulled together and lapped the circular sheet of metal will take on a concave shape and form a shade or reflector which will throw the light downwards. The overlapping edges of the reflector should be soldered or riveted to-

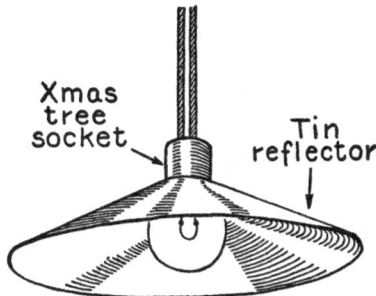

Fig. 272. — A Hanging Lamp fitted with Metal Reflector.

gether. The reflector is slipped over the base of the lamp, a small rubber or felt washer having been placed over the base next to the glass bulb so that the reflector will not break the lamp. The lamp is then screwed into a socket and allowed to hang downwards from a flexible conductor.

A very pretty effect can be secured by drilling the edges of a reflector full of small holes about three-sixteenths of an inch apart and then hanging short strings of beads from the holes. The beads should form a hanging fringe around the edge of

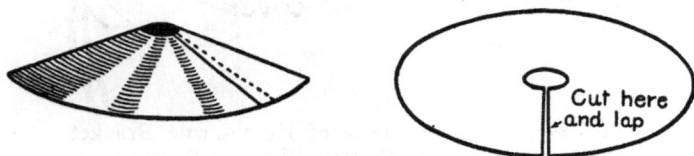

Cut here and lap

FIG. 273. — How the Reflector is made.

the reflector, and if they are of glass, a pleasing brilliancy is produced. Figure 273 shows how to make the reflector.

The batteries for a miniature lighting plant may be located in a closet, under a stairway, or in some other out-of-the-way place. Wires from there may be extended to various parts of the house, such as hallways, closets, the cellar stairs, over a shaving mirror in the bath-room or in any dark corner where a light is often temporarily needed. The wires can be run behind picture-mouldings or along the surbase and be almost entirely concealed.

With the aid of flashlight cells it is possible to construct a number of very useful electrical novelties and household articles in the shape of portable electric lamps, etc. Since flashlight cells are quite small they are only intended to operate small lamps. Only one lamp should be used on each battery at a time, and it should not be allowed to burn long. Some of these batteries will give many hours of intermittent service but if allowed to burn continuously would not light the lamp nearly so long. It is much the better plan to use them only for a few minutes at a time, and then turn the light off and allow the battery to recuperate.

An Electric Hand-Lantern is a very convenient device which is quite simple to make. It consists of a wooden box

FIG. 274. — A Home-made
Lantern.

large enough to receive three flashlight cells. The back of the box should open and close on hinges and be fastened with a hook so that the battery may be easily removed for renewal.

A three-and-one-half-volt tungsten lamp is mounted on the front of the lantern and connected with the battery and a switch so that the light can be turned on and off at will. The switch may be placed at the top of the box so that the fingers of the same hand used to carry the lantern may be used to turn the light on and off. The lantern is fitted with a leather strap at the top, to be convenient for carrying.

The Ruby Lantern is somewhat similar in arrangement to the hand-lantern and may be used both as a hand-lantern and a ruby light for developing photographs.

FIG. 275. — An Electric Ruby Lantern.

It consists of a wooden box to hold a three-cell dry battery, and is provided with a handle so that it may be easily carried. A switch by which to turn the lamp on and off is mounted on the side of the box.

The light is furnished by a three-and-one-half-volt tungsten lamp mounted on the front of an inclined wooden board arranged as shown in the illustration so as to throw the light downward. The sides and bottom of the box are grooved

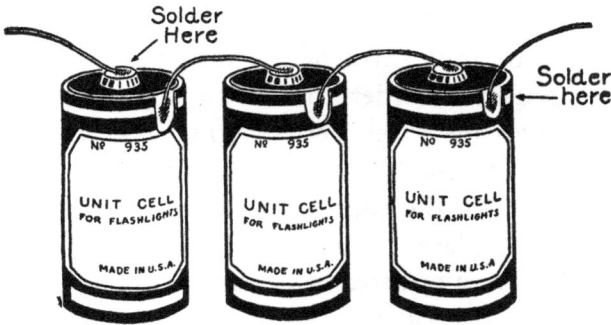

FIG. 276. — Three Flashlight Cells connected in series to
form a Battery for the Hand-Lantern.

near the front edges so that a piece of ruby glass may be inserted. Ruby glass for this purpose may be purchased at almost any store dealing in photographers' supplies.

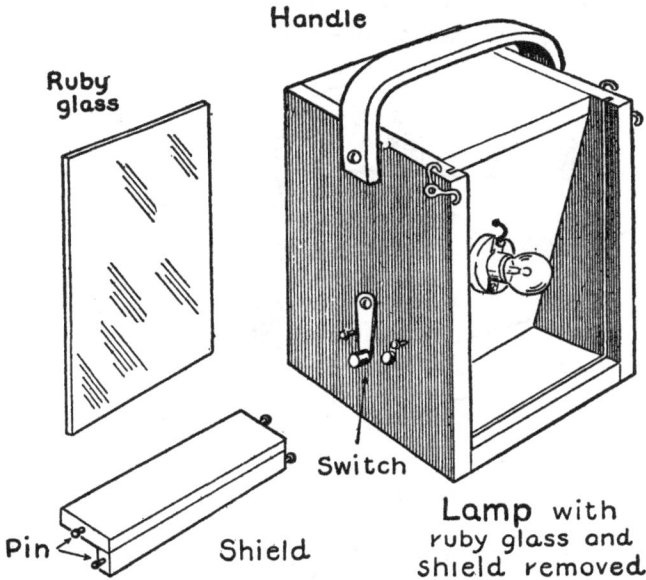

FIG. 277. — The Electric Ruby Lamp With Glass and
Shield Removed.

The top is provided with a shield which is fastened in position by means of four small hooks after the glass is in place. The shield is used in order to prevent any white light from escaping through the crack between the glass and the top of the box. A ruby lamp of this sort must be made absolutely "light-tight," so that the only light emitted is that which passes through the ruby glass. If any white light escapes it is liable to fog and spoil any pictures in process of development.

By removing the ruby glass and the shield, the light is changed into a hand-lantern. The back of the box should be made removable so that the battery can be replaced when worn out.

A Night-Light, arranged to shine on the face of the clock so that the time may be easily told during the night without inconvenience is shown in Figure 278.

It consists of a flat wooden box containing a three-cell

FIG. 278. — An Electric Night Light for telling the Time during the Night. The Switch is shown at the upper left.

dry battery and having a small three-and-one-half-volt tungsten lamp mounted on the top in the front with room for a clock to stand behind. The battery and the lamp are connected to a switch so that the light may be turned on and off. By attaching a long flexible wire and a push-button of the "pear-push" type it is possible to place the light on a table and run the wire with the push-button attached over to the bed so that one may see the time during the night without getting up. The bottom of the box should be made removable so that a new battery may be inserted when the old one is worn out.

The Watch-Light is in many ways similar to the clock light just described—but is smaller. It consists of a box just large enough to receive a three-cell flashlight battery. A piece of

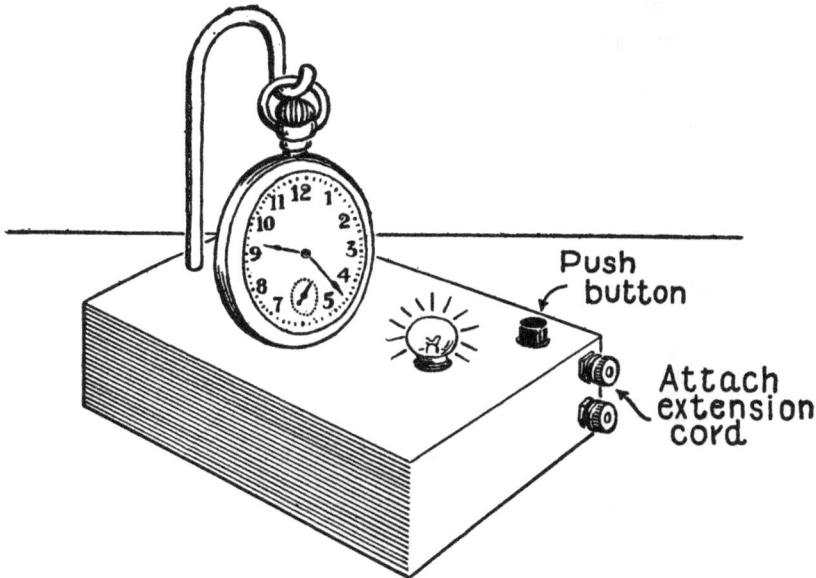

Push
button

Attach
extension
cord

FIG. 279. — A Watch-Light. The Battery is concealed in the Wooden Base.

brass rod is bent into the form of a hook or crane from which to suspend the watch.

The light is supplied by a three-and-one-half-volt tungsten

flashlight bulb mounted on the top of the box in front of the watch. If desirable, the light may be fitted with a small shade or reflector so that it shines only on the dial and not in the eyes. The figures on the face of the timepiece can then be seen much more plainly.

The lamp is mounted in a small wooden socket or a pin-socket passing through a hole in the top of the box, so that the wires are concealed. A small push-button is located in one of the forward corners of the box, so that when it is pressed the lamp will light. Two small binding-posts mounted at the lower right-hand corner of the box are connected directly across the terminals of the switch, so that a flexible wire and a push-button can be connected, and the light operated from a distance.

An Electric Scarf-Pin can be made by almost any boy who is skillful with a pocket-knife. The material from which the pin is made may be a piece of bone, ivory, or meerschaum.

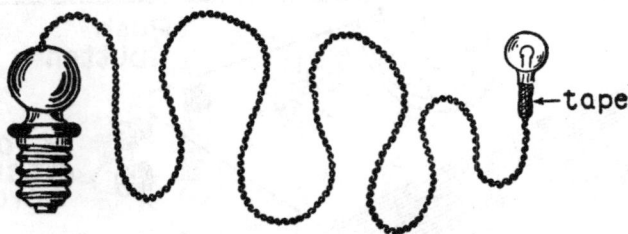

FIG. 280. — A "Pea" Lamp Attached to a Flexible Wire and a Plug.

It is carved into shape with the sharp point of a penknife and may be made to represent a skull, dog's head, an owl, or some other simple figure. The inside is hollowed out to receive a "pea" lamp. Pea lamps with a cord and a plug attached may be purchased from dealers in model supplies. The lamp is about one-eighth of an inch in diameter. The eyes, nose and mouth of the figure are pierced with small holes, so that when the lamp is lighted the light will show through the holes. The figure should be carved down thin enough to be translucent and light up nicely.

A large pin is cemented or otherwise fastened to the back of the figure so that it can be placed on the necktie or the lapel of the coat. The lamp is removed from the socket of a small flashlight and the plug attached to the pea lamp screwed into its place. The pea lamp is inserted inside the figure and bound

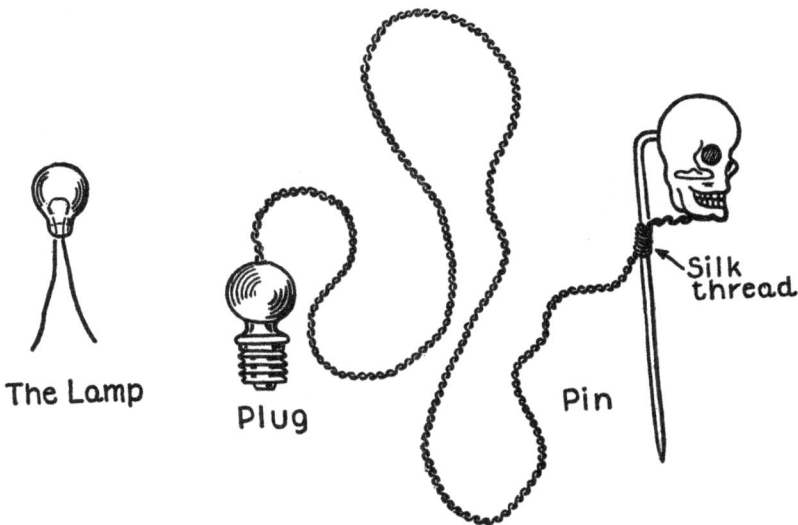

The Lamp

Plug

Pin

Silk thread

Fig. 281. — An Electric Scarf-Pin can be made by any Boy who is skillful with a Pocket-knife.

in place with some silk thread. Then when the button is pressed on the flashlight case, the pin will light up and tiny beams of light will shoot from the eyes, nose and mouth of the figure.

The drawings show how to carve a skull scarf-pin. It is made from a cylindrical piece of bone about five-eighths of an inch long and three-eighths of an inch in diameter. The first

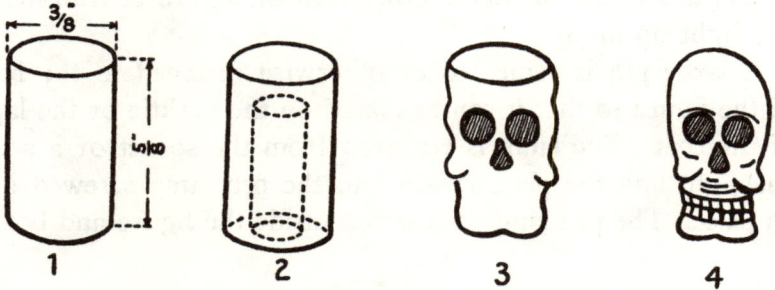

FIG. 282.—Four Steps in Carving a Skull Scarf-Pin. 1. The Bone. 2. Hole Drilled in Base. 3. Roughed Out. 4. Finished.

operation is to drill a hole three-eighths of an inch deep into the bottom. The hole should be large enough in diameter to pass the pea lamp.

Chapter XXI

MISCELLANEOUS ELECTRICAL APPARATUS

How Electricity May Be Generated from Heat

For the past century there has been a constant endeavor on the part of many scientists and inventors to "harness the sunlight." The energy which comes every day to our planet is incalculable. The energy consumed in the sun and thrown off in the form of heat is so great that it makes any earthly thing seem infinitesimal. We can only feel the heat from a large bonfire a few feet away, yet the sun's heat travels 90,000,000 miles before it reaches us, and even then our planet is receiving only the smallest fractional part of the total amount radiated.

Dr. Langley of the Smithsonian Institute estimated that all the coal in the State of Pennsylvania would be used in a fraction of a second if it were sent to the sun to supply energy.

Perhaps, some day in the future, electric locomotives will haul their steel cars swiftly from city to city by means of electricity, generated with "sun power." Perhaps energy from the same source will heat our dwellings and furnish us light and power.

This is not an idle dream, but may some day be an actuality. It has already been carried out to some extent. Several inventors have succeeded in making a device for generating electricity from sun energy.

One of these consists of a large frame, in appearance very much like a window. The glass panes are made of violet glass, behind which are many hundred little metallic plugs. The sun's heat, imprisoned by the violet glass, acts on the plugs to produce electricity. One of these generators exposed to the sun for ten hours will charge a storage battery and produce enough current to run 30 large tungsten lamps for three days.

The principle upon which the apparatus works was discovered by a scientist named Seebeck, in 1822. He succeeded in producing a current of electricity by heating the points of contact between two dissimiliar metals.

Any boy can make a similar apparatus, which, while not giving enough current for any practical purpose, will serve as an exceedingly interesting and instructive experiment.

Cut forty or fifty pieces of No. 16 B. & S. gauge German silver wire into five-inch pieces. Iron wire may be used in-

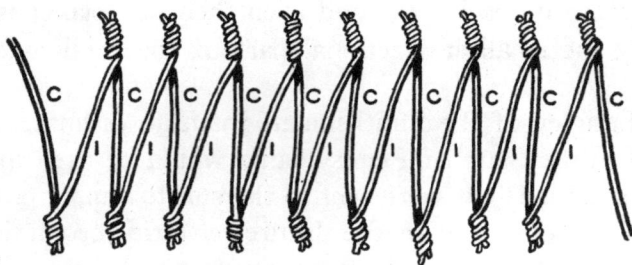

Fig. 283. — How the Copper Wires *C* and the German Silver Wires *I* are Twisted Together in Pairs.

stead of German silver but is not as satisfactory. Cut an equal number of similar pieces of copper wire, and twist each German silver wire firmly together with one of copper so as to form a zig-zag arrangement as in Figure 283.

Next make two wooden rings about four inches in diameter by cutting them out of a pine board. Place the wires on one of the rings in the manner shown in Figure 284. Place the second ring on top and clamp it down by means of two or three screws.

The inner junctures of the wires must not touch each other. The outer ends should be bent out straight and be spaced

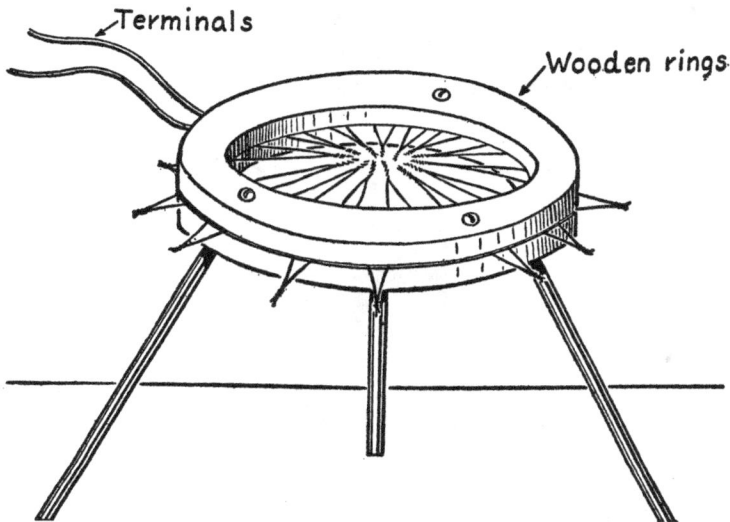

FIG. 284. — Complete Thermopile. An Alcohol Lamp Should Be Lighted and Placed So That the Flame Heats the Inside Ends of the Wires in the Center of the Wooden Ring.

equidistantly. The ring should be supported by three iron rods or legs. The two terminals of the *thermopile,* as the instrument is called, should be connected to binding-posts.

Place a small alcohol lamp or Bunsen burner in the center, so that the flame will play on the inner junctures of the wires.

The current may be very easily detected by connecting the terminals to a telephone receiver or galvanometer. By making several thermopiles and connecting them in parallel, sufficient current can be obtained to light a small lamp.

How to Make a Reflectoscope

A reflectoscope is a very simple form of a "magic lantern" with which it is possible to show pictures from post-cards, photographs, etc. The ordinary magic lantern requires a transparent lantern slide, but the reflectoscope will make pictures from almost anything. The picture post-cards or the photographs that you have collected during your vacation may be thrown on a screen and magnified to three or four feet in di-

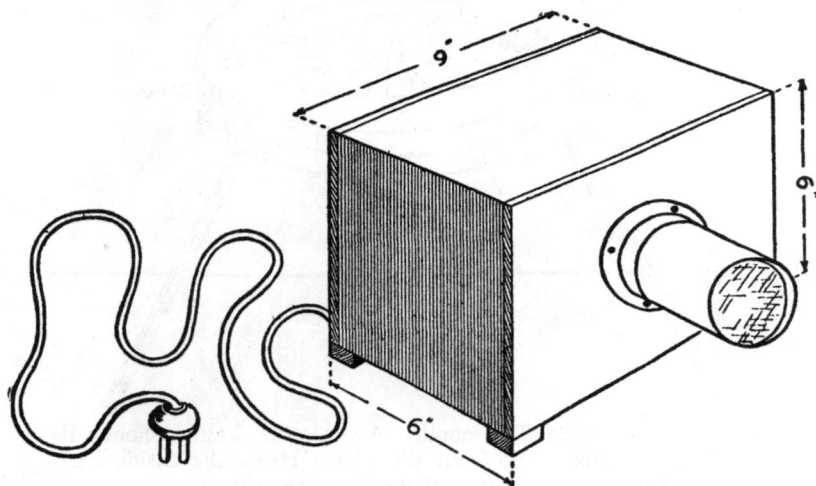

Fig. 285. — The Reflectoscope will project Pictures from Post-cards, Photographs, etc.

ameter. Illustrations clipped from a magazine or newspaper or an original sketch or painting will likewise show just as well. Everything is projected in its actual colors. If you put your watch in the back of the lantern, with the wheels and works exposed, it will show all the metallic colors and the parts in motion.

The reflectoscope consists of a rectangular box nine inches long, six inches wide, and six inches high outside. It may be built of sheet-iron or tin, but is most easily made from wood. Boards three-eighths of an inch thick are heavy enough. The methods of making an ordinary box are too simple to need description. The box or case in this instance, however, must be carefully made and be "light-tight," that is, as explained before, it must not contain any cracks or small holes which will allow light to escape if a lamp is placed inside.

A round hole from two and one-half to three inches in diameter is cut in the center of one of the faces of the box. The

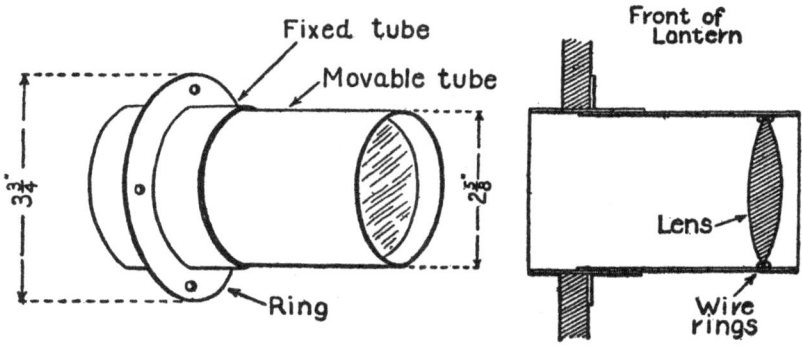

FIG. 286. — How the Lens Is Arranged and Mounted.

exact diameter cannot be given here because it will be determined by the lens which the experimenter is able to secure for his reflectoscope. Only one lens is required. It must be of the "double-convex" variety, and be from two and one-half to three inches in diameter.

A tube, six inches long and of the proper diameter to fit tightly around the lens, must be made by rolling up a piece of sheet-tin and soldering the edges together. This tube is the one labeled "movable tube" in the illustrations. A second tube, three inches long and of the proper diameter to just slip over the first tube, must also be made. A flat ring cut from stiff sheet-brass is soldered around the outside of this second tube, so that it may be fastened to the front of the case by three or four small screws in the manner shown. The hole in the front of the box should be only large enough to receive the tube.

The lens is held in position near one end of the movable tube by two strong wire rings. These rings should be made of

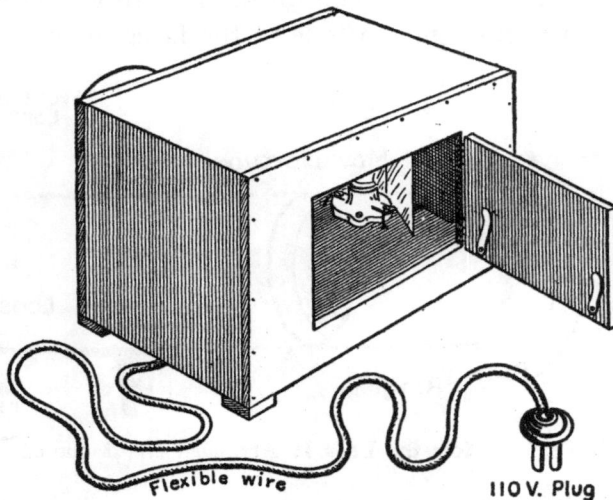

Flexible wire   110 V. Plug

FIG. 287. — A View of the Reflectoscope, From the Rear, Showing the Door etc.

wire that is heavy and rather springy, so that they will tend to open against the sides of the tube. It is a good plan to solder one of them in position, so that it cannot move, and then put in the lens. After the lens is in position, the second ring should be put in and pushed down against the lens. Do not attempt to put the lens in, however, until you are sure that the metal has cooled again after soldering, or it will be liable to crack. The back of the box contains an opening about four inches high and five and one-half inches long. The pictures that it is desired to project on the screen are held at this opening.

The light for the reflectoscope is supplied by two 75-watt tungsten lamps. The lamps fit into ordinary flat-base porcelain

FIG. 288. — A Socket for
Holding the Lamp.

receptacles. Two of these receptacles are required, one for each lamp.

The reflectors are made of tin, bent as shown. They are fastened in position behind the lamps by four small tabs.

FIG. 289. — The Reflector is
bent out of bright tin.

The interior of the reflectoscope must be painted a dead black by using a paint made by mixing lampblack and turpentine. The interior also includes the inside of the tin tubes.

The electric current is led into the lamps with a piece of flexible lamp-cord passing through a small hole in the case. An attachment-plug is fitted to the other end of the cord, so that it may be screwed into any convenient lamp-socket.

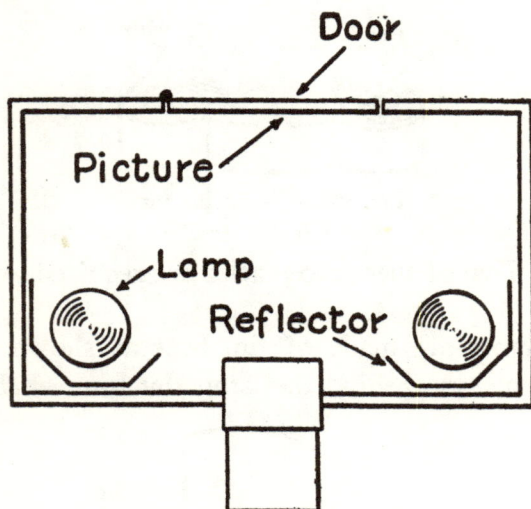

FIG. 290. — A View of the Reflectoscope With the cover Removed, Showing the Arrangement of the Lamps, etc.

The pictures should be shown in a dark room and projected on a smooth white sheet. The movable tube is slid back and forth until the picture on the screen becomes clear and distinct.

If four small feet, one at each corner, are attached to the bottom of the case, its appearance will be much improved.

Large pictures will tend to appear a little blurred at the corners. This is due to the lens and cannot be easily remedied.

How to Reduce the 110-V. Current So That It May Be Used for Experimenting

Oftentimes it is desirable to operate small electrical devices from the 110-v. lighting or power circuits. Alternating current of course can be reduced to the proper voltage by means of a small step-down transformer, such as that described in Chapter XIII. Another method is to use a *resistance*. The most suitable form of resistance for the young experimenter to use is a "lamp bank."

A lamp bank consists of a number of lamps connected in parallel, and arranged so that any device may be connected in series with it.

The lamps are set in "flat-base porcelain receptacles," mounted in a row upon a board and connected as shown in Figure 291.

The current from the power line enters through a switch and a fuse and then passes through the lamps before it reaches the device it is desired to operate. The switch is for the purpose of shutting the current on and off, while the fuse will "blow" in case too much current flows in the circuit due to an accidental short circuiting of any of the lamps.

The amount of current that passes through the circuit may be accurately controlled by the size and number of lamps used in the bank. The lamps may be screwed in or out and the current altered by one-quarter of an ampere at a time if desirable.

The lamps should be of the same voltage as the line upon which they are to be used. Each 25-watt lamp used will permit slightly less than one-quarter of an ampere to pass. Each 75-watt, 110-v. lamp will pass slightly more than one-half ampere.

AN INDUCTION MOTOR

An Induction Motor is a motor in which the currents in the armature windings are *induced*. An induction motor runs *without* any brushes, and the current from the power line is connected only to the field. The field might be likened to the primary of a transformer. The currents in the armature then constitute a secondary winding in which currents are induced in the same manner as in a transformer.

An induction motor will operate only on alternating current.

A small motor such as that shown in Figure 254, having a three-pole armature, is the best type to use in making an experimental induction motor.

Remove the brushes from the motor and bind a piece of bare copper wire around the commutator so that it short-circuits the segments.

A source of alternating current should then be connected to the terminals of the field coil. If you have a step-down transformer, use it for this purpose, but otherwise connect it in series with a lamp bank such as that just described.

Place a switch in the circuit so that the current may be turned on and off. Wind a string around the end of the armature shaft so that it may be revolved at high speed by pulling the string in somewhat the same manner that you would spin a top. When all is ready, give the string a sharp pull and immediately close the switch so that the alternating current flows into the field.

If this is done properly, the motor will continue to run at high speed, and furnish power if desirable.

Most of the alternating-current motors in everyday use for furnishing power for various purposes are induction motors. They are, however, self-starting, and provided with a hollow

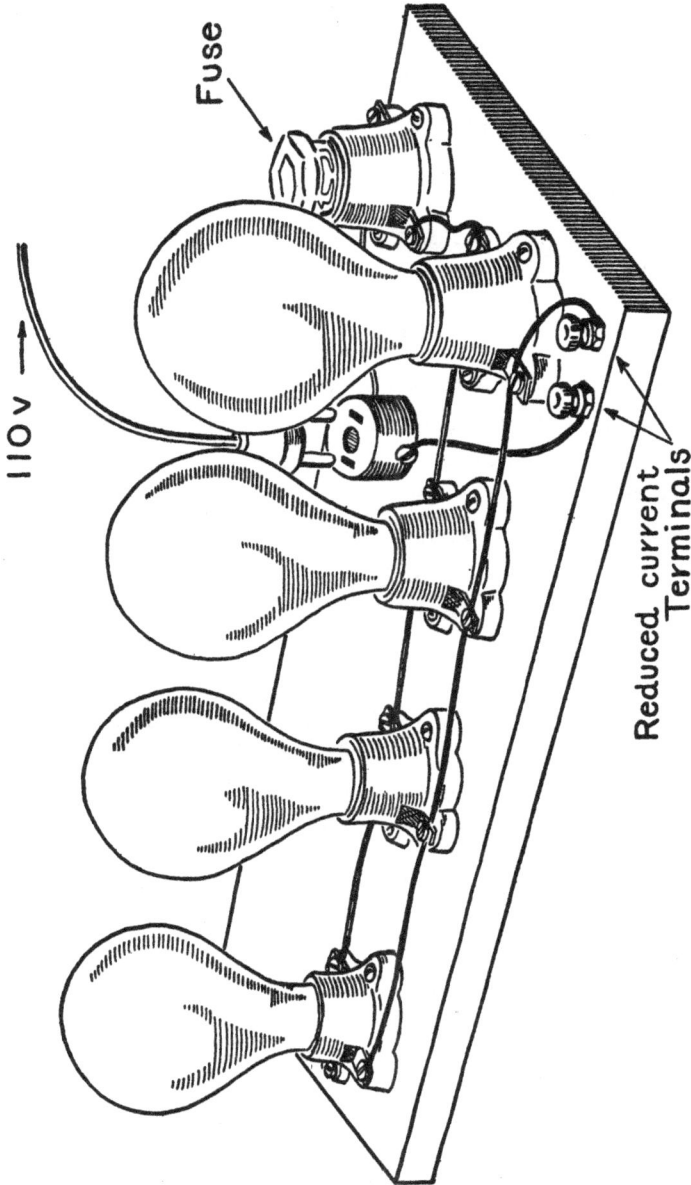

Fuse

110 v →

Reduced current
Terminals

Fig. 291. — A View of the Lamp Bank Showing How the Circuit is Arranged.

armature, which contains a centrifugal governor. When the motor is at rest or just starting, four brushes press against the commutator and divide the armature coils into four groups. After the motor has attained the proper speed, the governor is thrown out by centrifugal force and pushes the brushes away from the commutator, short-circuiting all the sections and making each coil a complete circuit of itself.

ELECTRO-PLATING

Water containing chemicals such as sulphate of copper, sulphuric acid, nitrate of nickel, nitrate of silver, or other metallic salts is a good conductor of electricity. Such a liquid is known as an *electrolyte*.

It has been explained in Chapter IV that chemical action may be used to produce electricity and that in the case of a cell such as that invented by Volta, the zinc electrode gradually wastes away and finally enters into solution in the sulphuric acid.

It is possible exactly to reverse this action and to produce what is known as *electrolysis*. If an electrolyte in which a metal has been "dissolved" is properly arranged so that a current of electricity may be passed through the solution, the metal will "plate out," or appear again upon one of the electrodes.

Electrolysis makes possible electro-plating and thousands of other exceedingly valuable and interesting chemical processes.

More than one-half of all the copper produced in the world is produced *electrolytically*.

Practically all plating with gold, silver, copper and nickel is accomplished with the aid of electricity.

These operations are carried out on a very large scale in

the various factories, but it is possible to reproduce them in any boy's workshop or laboratory, with very simple equipment.

The proper chemicals, a tank, and a battery are the only apparatus required. The current must be supplied by storage cells or a bichromate battery because the work will require five or six amperes for quite a long period.

A small rectangular glass jar will make a first class tank to hold the electrolyte.

The simplest electro-plating process, and the one that the experimenter should start with is copper-plating.

Fill the tank three-quarters full of pure water and then drop in some crystals of copper-sulphate until the liquid has a deep blue color and will dissolve no more.

Obtain two copper rods and lay them across the tank. Cut two pieces of sheet copper having a tongue at each of two corners so that they can be hung in the solution. Hang both of the sheets from one of the copper rods. Connect this rod to the *positive* pole of the battery. These sheets are known as the anodes.

Then if a piece of carbon, or some metallic object is hung from the other rod and connected to the *negative* pole of the battery, the electro-plating will commence. The apparatus should be allowed to run for about half an hour and then the object hung from the rod connected to the negative pole of the battery should be lifted out and examined. It will be found thickly coated with copper. It is absolutely necessary to have the poles of the battery connected in the manner stated, or no deposit of copper will take place.

Objects which are to be electro-plated must be free from all traces of oil or grease and absolutely clean in every respect, or the plating will not be uniform, because it will not stick to dirty spots. After all scale and dirt has been removed, dip the article to be plated in a lye solution, wash it and dry in clean saw-dust.

Such articles as keys, key-rings, tools, etc., can be prevented from rusting by coating with nickel.

Nickel-plating is very similar to copper-plating. Instead, however, of having two copper sheets suspended from the rod

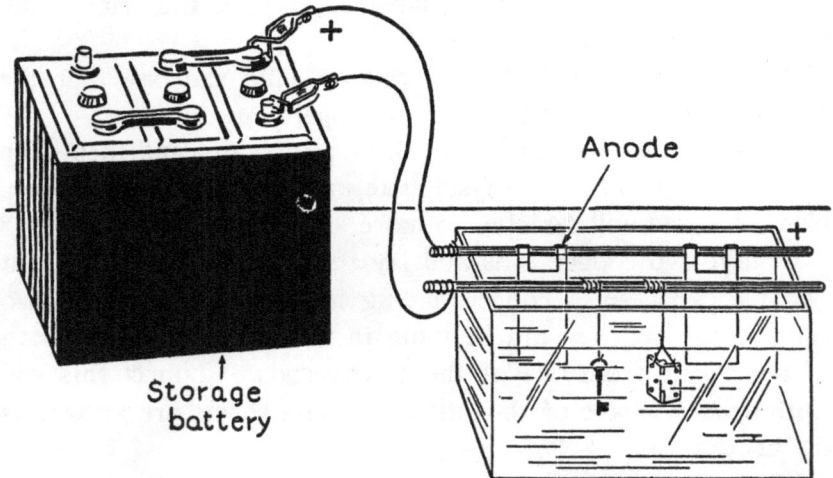

FIG. 292. — An Electro-plating Tank connected to a Storage Battery. Note that only one Cell of the Battery is in Use.

connected to the positive pole of the battery, they must be made of nickel.

The electrolyte is composed of one part of nickel-sulphate dissolved in twenty parts of water to which one part of sodium-bisulphate is added.

This mixture is placed in the tank instead of the copper-sulphate. The objects to be plated are hung from the copper rod connected to the negative pole of the battery.

When the nickel-plated articles are removed from the bath they will have a dull, white color known as "white nickel."

When white nickel is polished with a cloth wheel revolving at high speed, and known as a buffing-wheel, it will assume a high luster.

How to Make a Rheostat

It is often desirable to regulate the amount of current passing through a small lamp, motor, or other electrical device operated by a battery.

This is accomplished by inserting a resistance into the circuit. A rheostat is an arrangement for quickly altering the amount of resistance at will.

A simple rheostat is easily made by fitting a five-point switch such as that shown in Figure 293 with several coils of

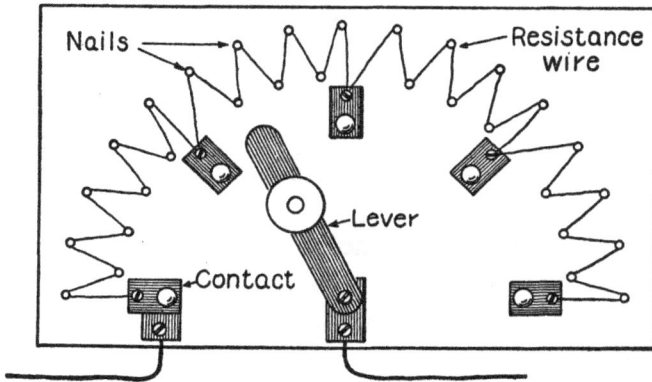

Fig. 293. — A Rheostat.

German-silver resistance wire. German silver has much more resistance than copper wire, and is used, therefore, because less will be required, and it will occupy a smaller space.

A five-point switch will serve satisfactorily in making a rheostat, but if a finer graduation of the resistance is desired it will be necessary to use one having more points.

Two lines of small wire nails are driven around the outside of the points, and a German-silver wire of No. 24 B. & S. gauge wound in zig-zag fashion around the nails from one point to the other.

The rheostat is placed in series with any device it is desirable to control. When the handle is on the point to the extreme left, the rheostat offers no resistance to the current. When the lever is placed on the second point, the current has to traverse the first section of the German-silver wire and will be appreciably affected. Moving the handle to the right will increase the resistance.

If the rheostat is connected to a motor, the speed can be increased or decreased by moving the lever back and forth.

In the same manner, the light from a small incandescent lamp may be dimmed or increased.

A CURRENT REVERSER OR POLE-CHANGING SWITCH

A motor in which the field is supplied by a permanent magnet can be reversed or caused to run in the opposite direction by merely changing the wires leading from the battery so that the current flows through the circuit in the opposite direction.

If the motor is provided with a field winding, however, the only way that the motor can be made to run in the opposite direction is by reversing the poles of either the armature or the field.

This is best accomplished by means of a pole-changing switch.

A pole-changing switch consists of a three-point switch having three points but only two levers. The levers are connected together by means of a strap so that they may be both moved by the same handle.

Such a switch may be made according to the design shown in Figure 294.

Motors such as those illustrated can be made to reverse by connecting to a pole-changing switch in the proper manner.

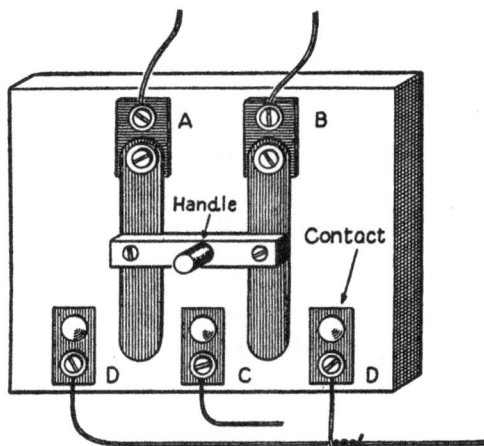

Fig. 294. — A Pole Changing Switch or Current Reverser. The Connecting Strip Is Pivoted So That the Handle Will Operate Both the Levers, *A* and *B*.

The two outside points or contacts (marked D and D) should both be connected to one of the brushes on the motor. The middle contact, C, is connected to the other brush.

One terminal of the field is connected to the battery. The other terminal of the field is connected to the lever, A. B connects to the other terminal of the battery.

When the switch handle is pushed to the left, the lever A should rest on the left-hand contact, D. The lever B should make contact with C. The motor will then run in one direction. If the handle is pushed to the right so that the levers

A and B make contact respectively with C and D (right-hand), the motor will reverse and run in the opposite direction.

How to Build a Tesla High-Frequency Coil

A Tesla high-frequency coil or transformer opens a field of wonderful possibilities for the amateur experimenter. Innumerable weird and fascinating experiments can be performed with its aid.

When a Leyden jar or a condenser discharges through a coil of wire, the spark which takes place does not consist simply of a single spark passing in one direction, as it appears to the

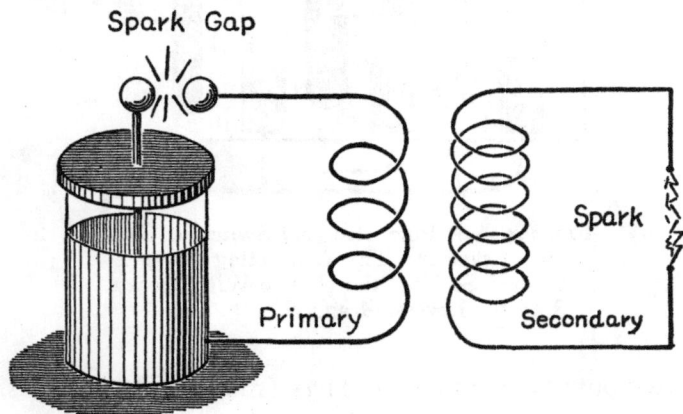

Fig. 295. — Illustrating the Principle of the Tesla Coil.

eye, but in reality is a number of separate sparks alternately passing in opposite directions. They take place so rapidly that the eye cannot distinguish them. The time during which the spark appears to pass may only be a fraction of a second, but during that short period the current may have oscillated back and forth several thousand times.

If the discharge from such a Leyden jar or a condenser is

passed through a coil of wire acting as a *primary,* and the primary is provided with a *secondary* coil containing a larger number of turns, the secondary will produce a peculiar current known as *high-frequency* electricity. High-frequency currents reverse their direction of flow or *alternate* from one hundred thousand to one million times a second.

High-frequency currents possess many curious properties. They travel only on the surface of wires and conductors. A hollow tube is just as good a conductor for high-frequency currents as a solid rod of the same diameter. High-frequency currents do not produce a shock. They do, however, burn in

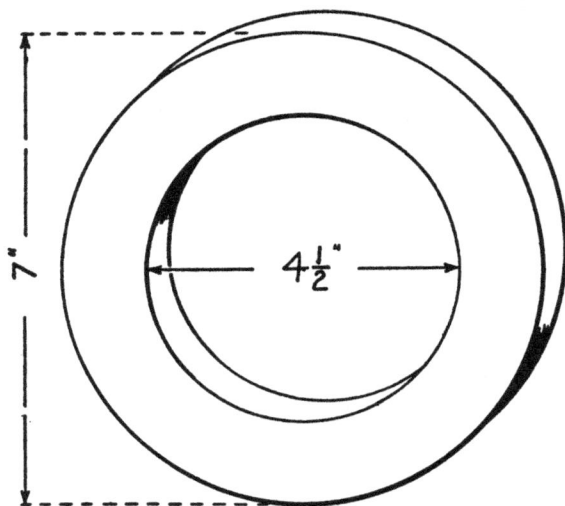

FIG. 296. — Details of the Wooden Rings Used As the Primary Heads.

some cases. If you hold a piece of metal in your hand you can take the shock from a high-frequency coil throwing a spark two or three feet long with scarcely any sensation save that of a slight warmth.

The Tesla coil described below is of a size best adapted for use with a two-inch or three-inch spark coil, or a small high-potential transformer. The purpose of the spark coil or the transformer is to charge the Leyden jars or condenser which discharge through the primary of the Tesla coil.

If the young experimenter wishes to make a Tesla coil which will be suited to a smaller spark coil, for instance, one capable of giving a one-inch spark, the dimensions of the Tesla coil herein described can be cut exactly in half. Instead of making the secondary twelve inches long and three inches in diameter, make it six inches long and one and one-half inches in diameter, etc.

The Primary consists of eight turns of No. 10 B. & S. gauge copper wire wound around a drum. The heads of the drum

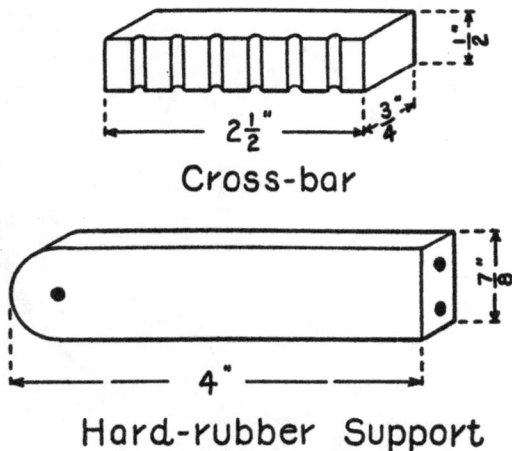

Cross-bar

Hard-rubber Support

FIG. 297. — The Cross-bars which support the Primary are Wood. The Secondary Supports are Hard-Rubber, Glass or Bakelite.

are wooden rings, seven inches in diameter and one-half inch thick. A circular hole four and one-half inches in diameter is cut in the center of each of the heads.

The cross bars are two and one-half inches long, three-quarters of an inch thick and one-half of an inch wide. Six cross bars are required. They are spaced at equal distances around the rings and fastened by means of a *brass* screw passing through the ring. When the drum is completed it should resemble a "Squirrel cage."

Small grooves are cut in the cross bars to accommodate the wire. The wires should pass around the drum in the form of a spiral and be spaced about five-sixteenths of an inch apart.

The ends of the wire should be fastened to binding-posts mounted on the heads.

The Secondary is a single layer of No. 26 B. & S. silk- or cotton-covered wire wound over a cardboard tube, twelve inches long and three inches in diameter.

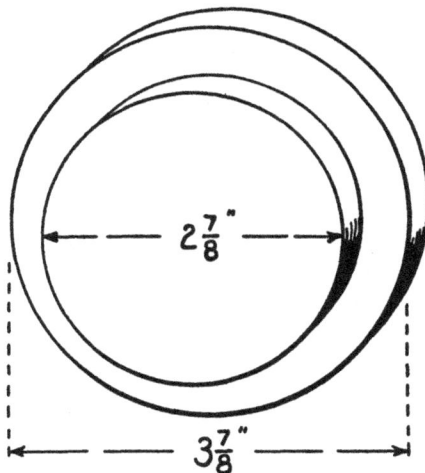

FIG. 298. — Wood Head for Secondary Tube.

The tube should be dried in an oven and then given a thick coat of shellac, both inside and out, before it is used. This

treatment will prevent it from shrinkage and avoid the possibility of having to rewind the tube in case the wire should
become loose.

The secondary is fitted with two circular wooden heads
just large enough to fit tightly into the tube, having a half-
inch flange, and an outside diameter of three and seven-eighths
inches.

The Base of the coil is fifteen inches long and six inches
wide and is made of wood.

The coil is assembled by placing the primary across the base

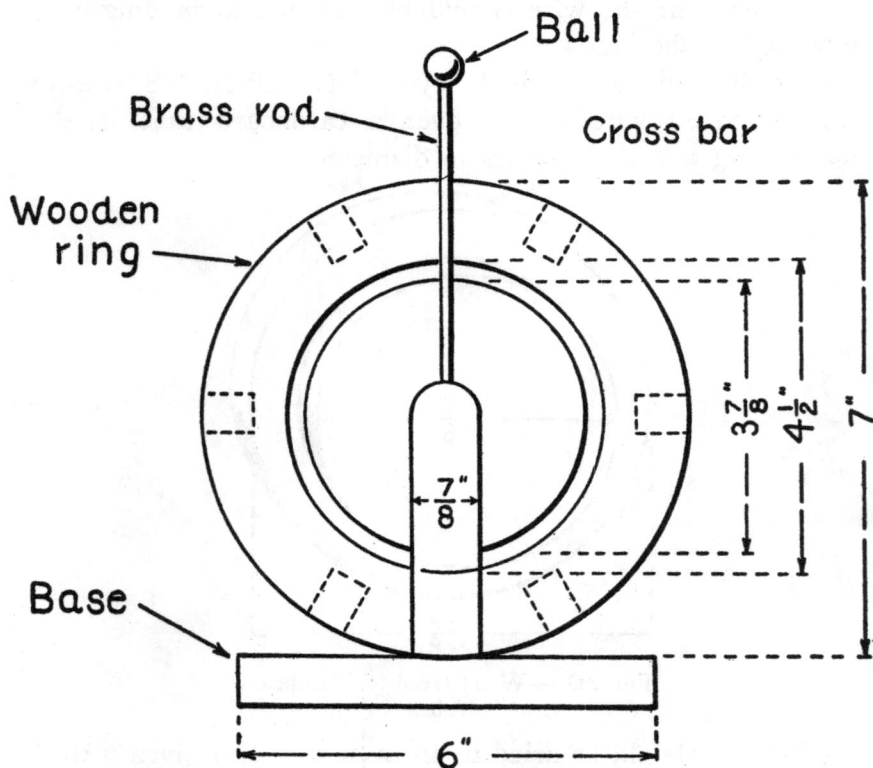

FIG. 299. — End View of the Complete Tesla Coil.

and exactly in the center. Two long wood-screws passing through the base and into the primary heads will hold it firmly in position.

The secondary is passed through the center of the primary and supported in that position by two hard rubber supports, four inches high, seven-eighths of an inch wide and one-half of an inch thick. A brass wood-screw is passed through the top part of each of the supports into the secondary heads so that a line drawn through the axis of the secondary will coincide with a similar line drawn through the axis of the primary.

The supports are made of hard rubber instead of wood, because the rubber has a greater insulating value than the wood. High-frequency currents are very hard to insulate, and wood does not usually offer sufficient insulation.

A brass rod, five inches long and having a small brass ball at one end, is mounted on the top of each of the hard rubber supports. The ends of the secondary winding are connected to the brass rods.

The lower end of each of the hard rubber supports is fastened to the base by means of a screw passing through the base into the support.

In order to operate the Tesla coil, the primary should be connected in series with a condenser and a spark-gap as shown

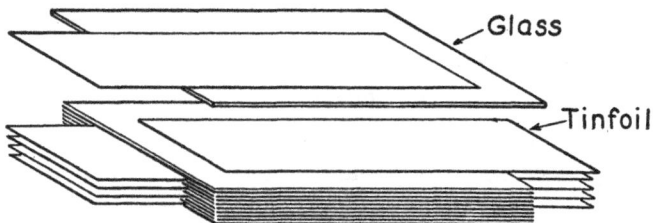

FIG. 300. — The Condenser is built of alternate Sheets of Tinfoil and Glass.

in Figure 301. The condenser may consist of a number of Leyden jars or of several glass plates coated with tinfoil. It is impossible to determine the number required ahead of time, because the length of the connecting wires, the spark-gap, etc., will have considerable influence upon the amount of condenser required. The condenser is connected directly across the secondary terminals of the spark coil.

When the spark coil is connected to a battery and set into operation, a snappy, white spark should jump across the spark-gap.

If the hand is brought close to one of the secondary terminals of the Tesla coil, a small reddish-purple spark will jump out to meet the finger.

Adjusting the spark-gap by changing its length and also altering the number of Leyden jars of condenser plates will probably increase the length of the high-frequency spark. It may be possible also to lengthen the spark by disconnecting one of the wires from the primary binding-posts on the Tesla coil and connecting the wire directly to one of any one of the turns forming the primary. In this way the number of turns in the primary is changed and the circuit is *tuned* in the same way that wireless apparatus is tuned by changing the number of turns in the tuning coil or helix.

The weird beauty of a Tesla coil is only evident when it is operated in the dark. The two wires leading from the secondary to the brass rods and the ball on the ends of the rods will give forth a peculiar *brush* discharge.

If you take a piece of metal in your hand and hold it near one of the secondary terminals, the brushing will increase. If you hold your hand near enough, a spark will jump on to the metal and into your body without your feeling the slightest sensation.

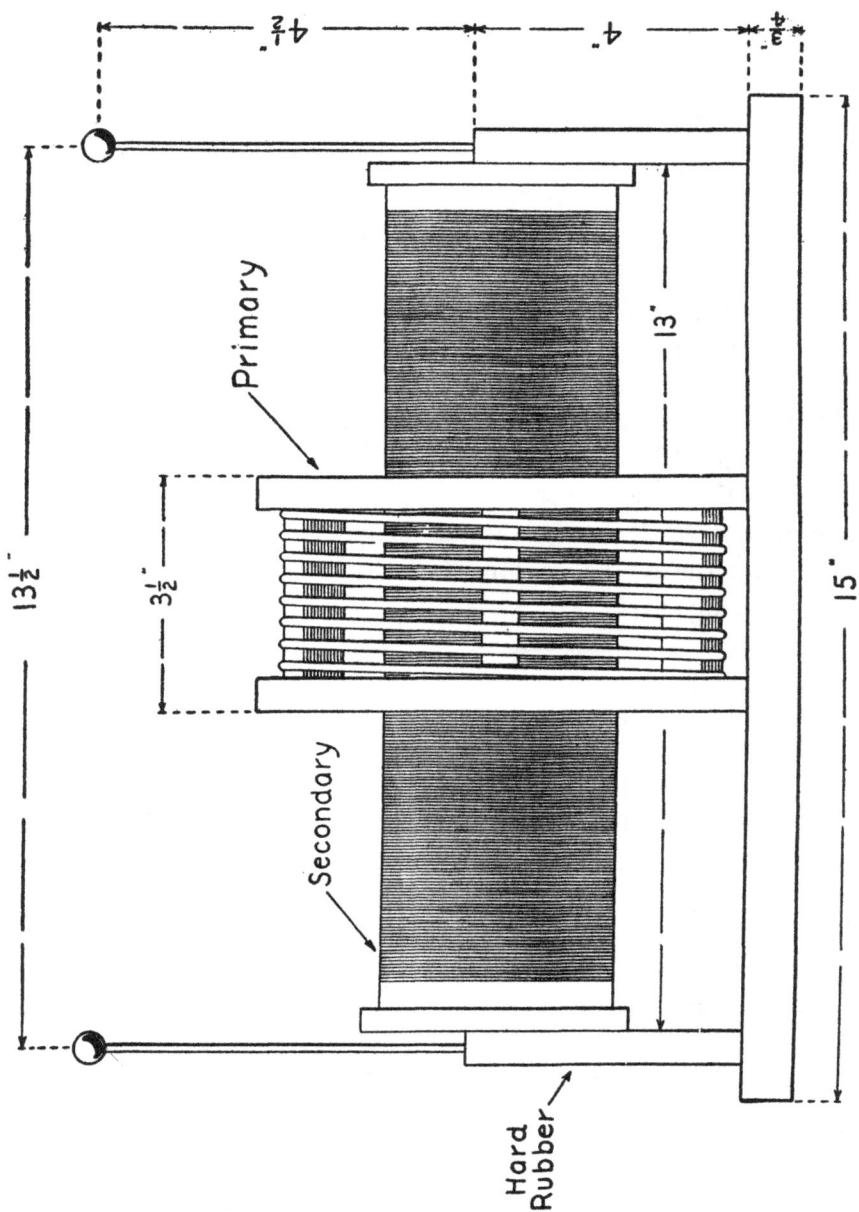

FIG. 301. — The Complete Tesla Coil.

387

If one of the secondary terminals of the Tesla coil is *grounded* by means of a wire connecting it to the primary, the brushing at the other terminal will increase considerably.

Make two rings out of copper wire. One of them should be six inches in diameter and the other one four inches in diameter. Place the small ring inside the large one and connect them to the secondary terminals. The two circles should be arranged so as to be *concentric,* that is, so that they have a common center.

The space between the two coils will be filled with a pretty brush discharge when the coil is in operation.

There are so many other experiments which may be performed with a Tesla coil that it is impossible even to think of describing them here, and the young experimenter wishing to continue the work further is advised to go to some library and consult the works of Nikola Tesla, wherein such experiments are fully explained.

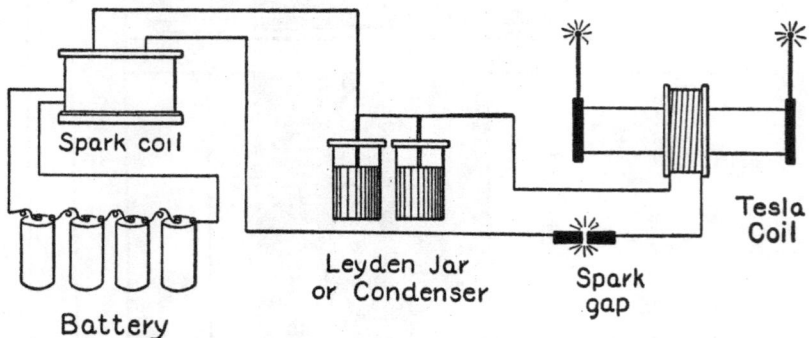

FIG. 302. —A Diagram Showing the Proper Method of Connecting a Tesla Coil.

PHOTOELECTRIC CELLS OR ELECTRIC EYES

One of the most interesting and amazing electrical devices is the photoelectric cell, or "electric eye," as it is popularly called. It made possible television, talking pictures, picture transmission by radio and wire. An electric eye will open doors, detect burglars, stop elevators, count and sort everything from marbles to men. It is faster and more sensitive than the human eye.

There are three types of photoelectric cells, the alkaline, electrolytic, and selenium. The alkaline cell is the type used most often in industry, in talking pictures, television apparatus, etc. It is the most sensitive but unfortunately cannot usually be made by the young experimenter. It is a factory made product and in appearance is somewhat like a radio tube.

Cells of the electrolytic type are not difficult to make, but they have the disadvantage of not being portable. The selenium cell is portable, small and not easily damaged if handled carefully. It is a satisfactory cell for the novice to make, if he is a careful mechanic.

Alexander Graham Bell, the famous scientist who invented the telephone, made the first practical selenium cells. They were built for use with the *photophone,* the first wireless telephone, an ingenious device in which words were carried between the transmitter and receiver *on a sunbeam.*

Selenium belongs to the same family of chemical elements as sulphur. It was discovered in 1817, by Berzelius, the Swedish chemist. Some of the elements have the peculiar property of existing in more than one form. Selenium is one of these. It can be a brown shapeless mass or exist in a crystalline form. Crystalline selenium has the strange property of being a conductor of electricity when exposed to light and almost an insulator in darkness.

A selenium cell usually consists of two electrodes, the space between the electrodes being filled with selenium. A current can flow from one electrode to the other, only by passing through the selenium. The resistance of the selenium to the flow of current will depend upon and vary with the intensity of the light shining on the selenium.

How to Make a Selenium Cell. The form of selenium cell easiest to build consists of a small glass tube about three-eighths of an inch in diameter and two inches long. Around this wind two No. 30 enameled copper wires parallel to one another. The two wires are temporarily tied to the tube with a piece of thread. Care must be taken to wind the wires evenly and smoothly, just as thread is wound on a spool. The tube should be wound in a single layer to within one-eighth inch of

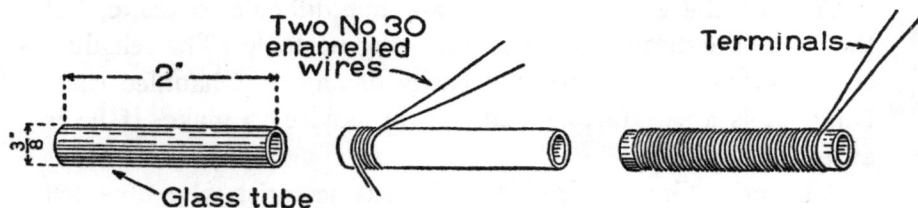

FIG. 303. — The Selenium Cell consists of two parallel Wires wound on a Glass Tube and coated with Selenium.

each end. When the end of the winding is reached, bind the loose ends firmly with thread so that the wire will not unwind. Then scrape the enamel from the outside surface of the wire wound around the tube, using a piece of very fine sandpaper or emery cloth for the purpose. Blow the dust off the winding and test the windings by placing in series with a telephone receiver and a battery to see whether or not the parallel wires are short-circuited. If no sharp click is heard in the receiver, the winding is not short-circuited and you can proceed with the next operation. If there is a short-circuit it is necessary to rewind the tube, using new wire and scraping the enamel from the outside surface just as before.

It is not always possible for the novice to build a successful selenium cell at the first attempt. A certain amount of skill is required. It is a good plan to prepare five or six windings. Although only one may prove to be satisfactory, you will still be well repaid for your time and trouble.

When the windings are completed, the cell is ready to be given a thin coating of selenium and treated to an annealing

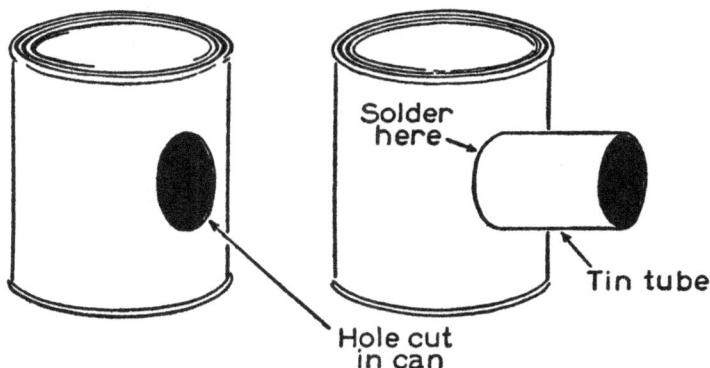

FIG. 304. — A Simple Method of making a Shield for the Selenium Cell or the Light Source is to cut a hole in the Tin Can and solder a piece of Tin Tubing over the hole.

process in a warm oven until the sensitive crystalline state is reached.

Buy a stick of selenium in the vitreous form from a chemical supply house. You should be able to secure a piece two or three inches long and about three-eighths of an inch in diameter for 25 to 50 cents.

Place the winding upon a small piece of sheet iron held *above* the flame of an alcohol lamp or Bunsen burner where it will be heated gently. When it is warm enough to melt the selenium, rub the latter over the winding until it is completely covered with a thin coating. If the winding is too hot, the selenium will not spread properly but will collect in drops. If too cold, the coating will be too thick. You must learn by the "trial and error" method just how much heat is required in order to melt the selenium properly.

You will need a small oven for the annealing process. You can make one from a tin can by fitting it with a shelf upon which to rest the cell. Place this oven over a Bunsen burner or alcohol lamp having a small flame so that a low heat is provided. Under no circumstances use enough heat to melt the selenium.

When the cell is put in the oven the selenium will soon assume an appearance like dull graphite. This is an indication that crystallization is taking place. If crystallization does not begin within ten minutes, increase the heat slightly. When crystallization has been completed (about 5 to 10 minutes after it begins) turn the heat up very gradually. You will have to watch carefully to insure good results. Increase the heat very slowly until the selenium just commences to show signs of melting. Remove the cell from the oven instantly. Turn the heat down again to a point slightly below the point at which the selenium commenced to melt and place the cell back in the oven again. Let it remain there for two or three hours. Then turn the heat down again and after fifteen minutes extinguish the flame. When the oven is cool the cell should be removed.

If you have had good luck and have followed instructions, the cell will probably have a resistance of 25,000 to 75,000 ohms in total darkness; but in the light, the resistance will drop sufficiently to close a sensitive relay connected in series with ten or twelve dry cells.

A galvanometer relay is the most sensitive but a telephone relay wound to a resistance of 3,000 ohms and requiring only four milliamperes to operate will prove satisfactory for ordinary demonstrations of the selenium cell.

The cell should be enclosed in a box or a tin can so that it is shielded from all light except that which comes through a

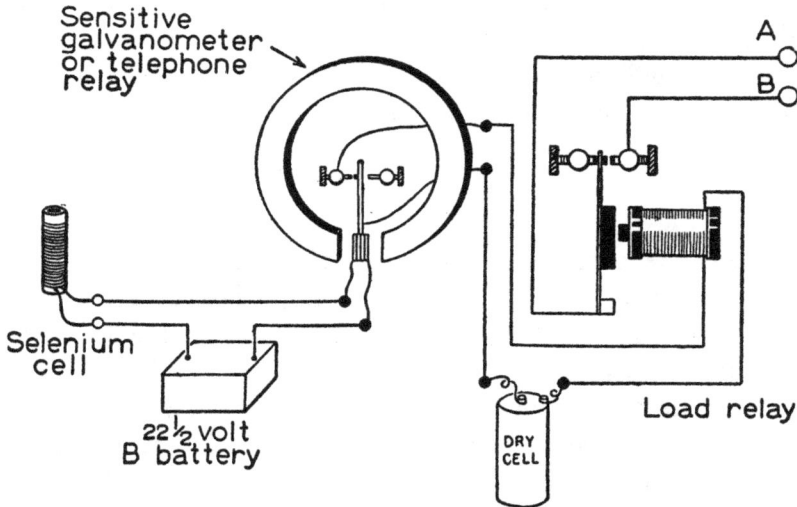

FIG. 305. — Diagram showing how to connect a Selenium Cell to a Galvanometer relay. An ordinary auto-horn Relay may be used as a load Relay.

hole in the side. The hole should be about two inches in diameter and provided with a shield so that only rays direct from the source of light can enter.

The source of light can be a 50 watt bulb enclosed in a can with a hole in the side and fitted with a shield. If the beam of light is aimed at the selenium cell any object passing between will interrupt the light and change the resistance of the cell so that the relay will open.

By changing the contact point on the relay from one side to the other, you can arrange so that the relay closes a circuit when a light is flashed upon the cell or closes it when the light is interrupted.

You can arrange your selenium cell to ring a bell when the sun rises or ring one when it sets. You can easily set up an efficient burglar alarm so that a bell will ring as soon as a light ray is intercepted. If the selenium cell is used to close a circuit in which more than an ampere is required, procure a power relay and connect it to the sensitive relay. An automobile horn relay will be satisfactory for this purpose. Its use will avoid burning the contact points of the sensitive relay.

Conclusion

Unless the average boy has materially changed his habits, in recent years, it matters not what the preface of a book may contain, for it will be unceremoniously skipped with hardly more than a passing glance. With this in mind, the author has tried to "steal a march" on you, and instead of writing a longer preface, and including some material which might properly belong in that place, has added it here in the nature of a conclusion, thinking that you would be more likely to read it last than first.

Some time ago, when in search for something that might be described in this book, I thought of some old boxes into which my things had been packed when I had dismantled my workshop before going away to college. They had been undisturbed for a number of years and I had almost forgotten where they had been put. At last a large box was unearthed from amongst a lot of dusty furniture put away in the attic. I pried the cover off and took the things out one by one and laid them on the floor. Here were galvanometers, microphones, switches, telegraph keys, sounders, relays and other things too numerous to mention. I was considerably amused and interested in the manner in which bolts, screws, pieces of curtain rod, sheet-iron, brass and other things had been taken to form various parts of the instruments. The binding-posts had almost in every case seen service as such on dry cells before they came into my hands. The only parts that it had been necessary to buy were a few round-headed brass screws and the wire which formed the magnets. In several instances, the latter were made so that they might be easily removed and mounted upon another instrument. The magnets on the telegraph sounder could be removed and fitted to form part of an electric engine or motor.

One particular thing which struck me very forcibly was the lack of finish and the crudeness which most of the instruments showed.

Of course it was impossible to avoid the clumsy appearance which the metal parts possessed, since they were not originally made for the part that they were playing, but I wished that I had taken a little more care to true up things properly or to smooth and varnish the wood, or that I had removed the tool-marks and dents from the metal work by a little filing.

If I had done so, I should now be distinctly proud of my work. That is not to say that I am in the least ashamed of it, for my old traps certainly served their purpose well, even if they were not ornamental and were better back in their box. Perhaps I might be excused for failing in this part of the work through lack of proper tools, and also because at that time there were no magazines or books published which explained how to do such things, and when I built my first tuning coils and detectors nothing on that subject had ever been published. I had to work out such problems for myself, and gave more thought to the principles upon which the instruments operated than to their actual construction.

The lads who read this book have the advantage of instructions showing how to build apparatus that has actually been built and tested. You know what size of wire to use and will not have to find it out for yourself. For that reason you ought to be able to give more time to the construction of such things. The purpose of this conclusion is simply a plea for better work. The American boy is usually careless in this regard. He often commences to build something and then, growing tired before it is finished, lays it aside only to forget it and undertake something else. *Finish whatever you undertake.* The principle is a good one. Remember also that care with the little details is what insures success in the whole.

If in carrying out your work, you get an idea, do not hesitate to try it. A good idea never refused to be developed. It is not necessary to stick absolutely to the directions that I have given. They will insure success if followed, but if you think you can make an improvement, do so.

Of course, such a book as this cannot, in the nature of things, be exhaustive, nor is it desirable, in one sense, that it should be.

The principle in mind has been to produce a work which would stimulate the inventive faculties in boys, and to guide them until face to face with those practical emergencies in which no book can be of any assistance but which must be overcome by common sense and the exercise of personal ingenuity.

The book is not free from technical terms or phrases, because certain of those terms have a value and an every-day use which are a benefit to the young experimenter who understands them.

Any subject treated in the various chapters of the "Boy Electrician" can be developed far beyond that point to which I have taken it. For instance, the railroad system could be fitted with electric signals, drawbridges, and a number of other devices.

Many new ideas suggest themselves to the ready-witted American lad. I shall always be pleased to hear from any boy who builds any of the apparatus I have described, and, if possible, to receive photographs of the work. I should be glad to be of any assistance to such a lad, but remember that it required many hours to complete some of the drawings and text in this book, and therefore please do not write to ask how to build other apparatus not described herein. And, as the future years bring new inventions and discoveries, no one now knows but that, some day, perhaps I shall write another "Boy Electrician."

THE END

INDEX